Loath to Print

Loath
to Print

The Reluctant Scientific Author, 1500–1750

 Nicole Howard

Johns Hopkins University Press | *Baltimore*

© 2022 Johns Hopkins University Press
All rights reserved. Published 2022
Printed in the United States of America on acid-free paper
9 8 7 6 5 4 3 2 1

Johns Hopkins University Press
2715 North Charles Street
Baltimore, Maryland 21218-4363
www.press.jhu.edu

Library of Congress Cataloging-in-Publication Data
Names: Howard, Nicole, author.
Title: Loath to print : the reluctant scientific author, 1500–1750 /
    Nicole Howard.
Description: Baltimore : Johns Hopkins University Press, 2022. | Includes
    bibliographical references and index.
Identifiers: LCCN 2021028402 | ISBN 9781421443683 (hardcover ; acid-free
    paper) | ISBN 9781421443690 (ebook)
Subjects: LCSH: Science publishing—History—16th century. | Science
    publishing—History—17th century. | Science publishing—History—18th
    century. | Scientists—Attitudes. | Technical writing—History. |
    Communication in science—History.
Classification: LCC Z286.S4 H7278 2022 | DDC 070.5—dc23/eng/20211105
LC record available at https://lccn.loc.gov/2021028402

A catalog record for this book is available from the British Library.

Special discounts are available for bulk purchases of this book. For more
information, please contact Special Sales at specialsales@jh.edu.

For my parents, Lynn and Jim Howard

Contents

Acknowledgments

This book is the product of a long gestation, and I am grateful to so many for helping me see it to completion. In the early stages of research, I was given invaluable advice by my friend and fellow early modernist Millie Gimmel, who patiently heard me out as I considered the shape of the argument. Panel presentations on aspects of the project delivered at the Renaissance Society of America, SHARP, and the History of Science Society conferences were also extremely useful. I especially want to thank Elizabeth Yale and an anonymous reviewer for their extensive and thoughtful comments on an early draft. Closer to home, I received feedback from the Columbia History of Science Group, which meets annually in the San Juan Islands. Known as the westernmost and soggiest gathering of historians of science, the meeting provided me with critical and constructive responses to my work-in-progress. Anita Guerrini was particularly helpful and supportive.

I am very much in debt to the many librarians and archivists who have helped this project along the way, including those from the Huntington Library, the David M. Rubenstein Rare Book and Manuscript Library at Duke University, and the Special Collections at the University of Leiden. A sabbatical award from Eastern Oregon University gave me the year I needed to write the book, and support from the Eastern Oregon University Foundation helped with procurement of images.

Finally, I owe a great deal to colleagues and friends who have encouraged me as the book evolved. Nico Bertoloni Meli has been a great mentor to me and has modeled the highest levels of historical scholarship. I am a better historian for having worked with him. Closer to home Ryan Dearinger has been a tremendously supportive colleague. I appreciate his intellectual stimulation, his encouragement, his dedication to the craft of history, and his willingness to work through anything over a beer.

My greatest debt is to my family. David Moyal has patiently watched me shepherd this book to publication, reading and commenting on multiple drafts, for which I am truly grateful. And my son Ephraim, who has spent half his life watching this book take shape, has become quite adept at giving mom time to work. He and his patience are a gift.

Loath to Print

"A Vast Ocean of Books"

Why should I, in so vast an ocean of books whereby the minds of the studious are bemuddled and vexed; of books of the more stupid sort whereby the common herd and fellows without a spark of talent are made intoxicated, crazy, puffed up; are led to write numerous books and profess themselves philosophers, physicians, mathematicians, and astrologers, the while ignoring and condemning men of learning: why I say should I add aught further to this confused world of writings.

—WILLIAM GILBERT, *DE MAGNETE* (1600)

In 1493, the German printer Anton Koberger published the *Liber Chronicarum*, a wide-ranging history of the Christian world written in Latin by the physician Hartmann Schedel. An example of early print and illustration techniques, it is one of the most famous typographic examples from the incunable period, the first fifty years of print in the West. In addition to the *Chronicle*'s description of various European and Near Eastern cities and its biblical genealogies, there are hundreds of portraits of historically influential individuals, from Greek gods to Assyrian kings to medieval scholars. Hercules, Socrates, Cleopatra, and the Venerable Bede are all depicted. Among these portraits is one of Nuremberg's most notable fifteenth-century residents, the astronomer and mathematician Johannes Müller (1436–1476), known to many as Regiomontanus. Indeed, while most individuals in the *Chronicle* are represented with interchangeable woodblocks, Müller's image is believed to have been personalized—he even holds an astrolabe—since he was known to both Schedel and the artist who worked on the book. Müller's likeness on the printed page of Nuremberg's most famous incunable serves as an appropriate starting point for an examination of the relationship between early modern print culture and the sciences.[1]

Regiomontanus received his early education in Leipzig and completed his university work in Vienna, then a center for astronomical studies. In particular, he studied under Georg von Peuerbach (1423–1461), whose work on plan-

vnd3 wiſchen Alphonſo dem kön
tochter ſchützet. In dem nehſt dat
Iohānes võ Königſperg tugalie

Tu
niſchet
von fü
weißh
det. vn
achtpe
ſyñſch
dere di
mit rec
Sirto
de zere
nochm

Icolaus eſſenſis der ſich vm
nachfolgend gefangen vnd :

Regiomontanus from *Liber Chronicarum* (1493). *Courtesy of the Huntington Library.*

etary motion was well known. In 1457 Regiomontanus completed his degree, but it was an encounter three years later that proved transformative for the young astronomer. In 1460, the great humanist Cardinal Bessarion visited Vienna, met Regiomontanus, and invited him to Italy to work on a new translation of Ptolemy's *Almagest*, then the most important work in astronomy. From Bessarion Regiomontanus learned Greek and gained familiarity with a host of classical astronomical works. His travels also allowed him to locate and acquire a number of important manuscripts relating to ancient mathematics, from Euclid's *Elements* to Apollonius' *Conics* to the algebraic treatises of Diophantus.[2] His goal in obtaining these manuscripts was to make such classical works available to other mathematicians, just as he and Bessarion sought to do with Ptolemy. But at precisely the moment when twenty-four-year-old Müller was unearthing and copying valuable manuscripts in Italy, the new technology of print was spreading from Germany to the Italian Peninsula. In 1464, during the middle of Regiomontanus' visit, the first Italian printing press was established by German monks in Subiaco, near Rome. We do not know when or how Regiomontanus was made aware of the new technology, but he was quick to seize on it. This scholar, who had spent years collating and commenting on classical works for other mathematicians, recognized the power of print to exponentially increase his influence on astronomical and mathematical learning.

After moving to Nuremberg in 1471, Regiomontanus enlisted a patron's support to start his own press in an office on Kartäusergasse. Correspondence from that year lays out his ambitious plans to print all of the important astronomical books, and word of his private press spread quickly. A letter to Hartmann Schedel from his uncle alludes to the mystery of this new printing operation:

> This Master Johannes [Müller] has left Nuremberg, but with the intention of returning; since he keeps all of his work secret, he has told no one, or just a few, about his departure . . . They said that he has printed an edition of 1000 copies of the new calendar with the true paths of the sun and moon. There isn't anybody who has seen it, except for the type-setters . . . I hear that he is now having his type-setter print an almanac of all the planets from many years. Until now neither I nor anyone else has been able to enter his house; however, one of these days I'll gain entrance . . . and inspect his work; I will let you know if I find out anything about his work.[3]

Regiomontanus' press issued its first book in 1472, Georg von Peuerbach's *Theoricae Novae Planetarum* (New theories of the planets), which was quickly adopted for university courses. Other publications followed, including the ambitious 1474 *Tradelist*, a printed broadside detailing the many astronomical and mathematical works, both his own and those of others, that Regiomontanus planned to print.[4] Copies of the *Tradelist* were sent to universities as a kind of advertisement, but Regiomontanus' untimely death in 1476 brought the project to an abrupt halt. Nevertheless, in the brief time his press was operational, he introduced new bibliographic conventions especially suited to the sciences, including large illustrations of planetary models, printed geometric diagrams, and typeset astronomical tables.[5]

Regiomontanus is unique among scientific authors because his life perfectly spanned the pre- and post-printing era. He was able to see the potential of print technology immediately but was equally attuned to the ways that created potential risk. The care and fidelity he had demonstrated in copying manuscripts could not wane with the press, "for who does not realize that the admirable art of printing recently devised by our countrymen is as harmful to men if it multiplies erroneous works as it is useful when it publishes properly corrected editions?"[6] His brief time as a printer did, indeed, set the bar for future presses. It is not surprising, then, that Müller's likeness appears in the *Nuremberg Chronicle*. The man who made printing central to science, and made Nuremberg a center of scientific activity, quite appropriately captured himself in print.

In the history of science, the consensus has been that the invention of printing with moveable type was central to the rapid growth of scientific knowledge in the sixteenth and seventeenth centuries, a burst of advancement known as the Scientific Revolution. Prior to print, the exchange of ideas was typically confined to small communication networks or gatherings of intellectuals, often in universities, salons, or private homes. Astronomers, mathematicians, and natural philosophers shared manuscripts, but the medium itself, the laborious recording of observations and ideas on parchment, intrinsically limited both the pace of production and the number of viewers for a given work.[7] This scribal culture was also linked to orality. University professors lectured to students in the sciences, and natural philosophers conversed about the nature of matter, the structure of the universe, or the latest unproven theorem. Printing with moveable type fundamentally changed and multiplied those conversations. Certainly this transition was neither as sudden nor as seismic as previous histories have assumed. The so-called print triumphalists have been challenged by scholars who rightly emphasize the persistence of a contemporary oral and manuscript culture.[8] Pen, paper, and conversation hardly lost their efficacy as scientific tools, as essential to knowledge-making as the telescopes and air pumps that followed them. Nevertheless, one cannot deny that by the sixteenth century, the sheer number of print shops, estimated at around one thousand by the year 1500, and the amount of printed material they produced revolutionized the sciences.[9] By the early seventeenth century, until the Thirty Years' War disrupted the market, the Frankfurt Book Fair offered one thousand new publications annually from publishers in Switzerland, the Low Countries, France, England, and the Italian Peninsula.[10]

Though many great works of science appeared in the sixteenth century, changes had already begun in the incunable period. Johannes de Spira printed

Pliny's *Natural History* in Venice in 1469, Ptolemy's *Geography* was printed in Bologna in 1475, and Philippus Pincius produced a Latin edition of Galen in 1490. This was only the beginning. After 1500, the number of printed scientific works grew exponentially. The appearance of a comet in 1577, for example, generated more than one hundred printed accounts. Print runs of 250–500 copies of a book or pamphlet were not uncommon, and some runs went as high as 3,000, a staggering number considering the effort involved in production of a single medieval manuscript. The first edition of Copernicus' *De Revolutionibus* had an initial print run estimated at 500 copies, Galileo had 550 copies of his *Sidereus Nuncius* printed in 1610, and around 750 copies of Newton's *Principia Mathematica* came off the press in 1687.[11] By the seventeenth century, print shops such as the Elzevirs in the Low Countries or John Martyn in England, specializing in mathematics, natural philosophy, medicine, and natural history, began to emerge.

As print technology developed, scholars became increasingly attuned to the production of scientific works, both in terms of what they needed to publish and what they anticipated reading once it was published. Erasmus, knowing that a new edition of Galen's medical works was in production, nudged Venetian publisher Aldus Manutius with his observation that "scholars are waiting impatiently for [it]."[12] A complete five-volume edition of Galen's work finally appeared in 1525 to much acclaim among those who could afford to purchase it. It did not take long for the scholarly community, no longer content to share manuscripts and letters, to recalibrate its expectations based on the potential of print. Scientists still shared letters and exchanged ideas through various forms of manuscript media, but the potency of the printing press, the sheer access it afforded, was evident to everyone in the intellectual community.

Elizabeth Eisenstein offered the first substantive look at this revolutionary technology in her landmark 1979 book *The Printing Press as an Agent of Change*. In two dense volumes, she articulates the important role that printing played in advancing the sciences, and she demonstrates the ways that print culture expanded access to, and fostered debates about, such theories as the Copernican universe, the circulation of blood, and universal gravitation. Central to Eisenstein's argument is the notion that more books in circulation allowed for a wider dissemination of ideas.[13] In asserting this, she stresses the importance of a text's "fixity," the idea that print not only allowed for the rapid proliferation of hundreds or thousands of copies, but that those copies could be exactly alike, allowing scholars to converse across great distances about the same ideas, which appeared on the same page, with identical images. When first published, Eisenstein's argument was as important as it was innovative, and her ideas gave rise to numerous conversations about the history of print and the place of books in Western culture.

In the four decades since, historians of print culture have augmented, challenged, and modified Eisenstein's claims. Scholars have highlighted, for example, the incredible variability in printed works, even within the same print run, to remind us that we should not view print as a process that created perfect, identical copies.[14] On a case-by-case basis, we have learned that the world of print was messy and complex, hardly a paradigm of control and order. Sentences only partially printed were often completed later by a reader's hand, and entire leaves could be added to a finished work, sometimes in a different font, sometimes by a different printer. The printing press, we now know, did not produce anything near a monolithic book culture, as I will discuss below. Nevertheless, Eisenstein's argument holds true on the macro level. By the early sixteenth century, readers could benefit from literally "being on the same page." As historian David Wooton writes, for example, "Printing created a community of astronomers working on common problems with common methods and reaching agreed solutions."[15] Wooten and Eisenstein are, in some ways, echoing the enthusiasm for print we see among early modern figures themselves. Consider the astronomer Johannes Kepler, who noted in his 1606 *De Stella Nova*, "Every year, especially since 1563, the number of writings published in every field is greater than all those produced in the past thousand years. Through them there has today been created a new theology and a new jurisprudence; the Paracelsians have created medicine anew and the Copernicans have created astronomy anew. I really believe that at last the world is alive, indeed seething, and that the stimuli of these remarkable conjunctions did not act in vain."[16] And we need look no further than Kepler's collaborator, Tycho Brahe, for an example of print's power. A Danish astronomer with tremendous observational skills and enough money to purchase his own printing press, Tycho was connected to a network of astronomers and mathematicians across Europe, including Paul Wittich, an itinerant teacher from Wratislavia. Tycho, Wittich, and others worked tirelessly to understand and critique Copernican ideas about the universe by commenting on printed copies of his book and sharing annotations with students and each other.[17] They even considered printing copies of their notes within the Copernican text. Their collaboration, which led in part to Tycho's formation of a new structural model for the universe, is a perfect example of the synergy Eisenstein describes, one that could not have existed with nearly the same efficiency prior to the printing press.

I want to suggest, however, that the embrace of printing technology by the scientific community was not as fervent as it might appear at first glance. In looking at personal correspondence, dedications and prefaces, exchanges between scientists and printers, and perhaps most importantly, at plans for the dissemination of scientific books in this period, one is struck by the chorus of hesitation scientists expressed over publishing. Concerns included the potential for printers to introduce errors into their works, the prospect of having them pirated, and most pressing, the likelihood that they would be misread and misunderstood by an audience that was either ill prepared to negotiate the complexities of the mathematical or philosophical ideas, or actively hostile to them. Scientists welcomed debate in this period, but when provocative ideas, such as the relocation of the earth away from the center of the universe or the prospect of life on other planets, were presented to a wider public, odds were that the ensuing debate would be grounded in neither logic nor mathematics but would hinge on principles of ideology and faith. Francis Bacon, among the period's most fervent champions of the printing press, lauded it as one of the crowning technological achievements of the age. Yet even he conceded it was not the technology itself that mattered: "For I care little about the mechanical arts themselves: only about those things which they contribute to the equipment of philosophy."[18] Bacon praised the printing press but only insofar as it promoted natural knowledge, a posture natural philosophers and their scientific peers embraced.

Such reservations bring to mind a particular Renaissance frontispiece from an edition of Edmund Spenser's *Faerie Queen*. This is hardly a scientific work, but the emblem on the 1617 title page provides a useful metaphor for a conversation about early modern books and their readers. In the emblem, a boar (sometimes thought to be a porcupine) backs away from a rosemary bush, the emblem of secrecy, beneath the motto *spiro non tibi*, or "I breathe out [sweet scents] but not for thee."[19] As Renaissance scholars have shown, by its nature the emblem serves those who can read the Latin motto. "Either the emblem means nothing to the reader or it refers to someone *else*; it is impossible to see oneself as the pig."[20] This emblem was not uncommon in the early modern period, appearing on the title pages of works by Erasmus, Sidney, Spenser, Machiavelli, and in the emblem encyclopedia of Joachim Camerarius in the late sixteenth century.

Frontispiece from Edmund Spencer's *Faerie Queen* (1617). *Courtesy of the Huntington Library.*

It also represents a sentiment that is central to this study: the idea that the material in certain books, while worthy of both study and praise, has a specific intended audience. As the rosemary bush is unfit for the crude sensibilities of the boar, a scientific treatise, according to many authors I examine, had no business in the hands of an unlearned reader. Although the motto *spiro non tibi* does not appear on the title page of a work of natural philosophy or astronomy, it gives expression to the attitudes conveyed throughout the correspondence and published work of natural philosophers and their peers. If knowledge of nature was the rosemary bush, early modern scientists approached print with a genuine worry about boars. Such concern led scientific authors to carefully consider the ways they wanted to use print and to develop strategies for engaging with or, conversely, circumventing the broad readership that print afforded. This book, then, is a study of the attitudes and tactics that authors from the sciences brought to bear on their work with the printing press.

To understand how authors thought about their readers requires more than attention to printed books, pamphlets, and images. It demands that we trace the work's path from the shop where it was created out into the public and private spheres where it was received. Thus, newer approaches in book history are important. Chief among these, and integral to this project, is the role of reader reception. By studying the way readers interacted with printed materials, their methods of annotation, their comparison of multiple sources, or their correspondence with others, historians can begin to understand the refractive nature of print's reception. If an author's ideas tended to enter the print shop in homogeneous fashion, the press allowed for a kind of prismatic effect among readers. The resulting spectrum was witnessed by authors, who responded to readers' agency in a variety of ways. These authorial responses are at the core of this project.

I begin chapter 1 by considering where the printing press fit into the larger conversation about creating and sharing natural knowledge. This requires a survey of attitudes scientists held toward print, which ranged from complete mistrust to a wholesale embrace of the press. I then consider how those attitudes manifested in publishing decisions. Some authors embraced print and found creative ways to use it to their advantage, professionally, intellectually, and monetarily. For them, the printing press was the primary vehicle for establishing their ideas and affirming their contributions to a scientific field. There was no liability in widespread circulation. There are particularly good examples of such authors in medicine and domestic sciences, such as husbandry, cooking, forestry, and natural history. Scientists from other fields, particularly mathematics, natural philosophy, and astronomy, approached print with a caution that sometimes resulted in a complete unwillingness to publish at all. Pierre de Fermat is a case in point. The French mathematician was not inclined to print his works but rather chose to circulate his papers among particular individuals. By doing this he created the intellectual exchange he needed without the burden of worrying about his audience. Fermat was content to write for himself and his peers. A generation later, Fermat's attitude was echoed by Isaac Newton. In 1676, eleven years before he would publish the famous *Principia Mathematica*, Newton complained to Henry Oldenburg that "a man must either resolve to put out nothing new or to become a slave to defend it."[21] Unlike Fermat, however, Newton could not completely disengage from the pressure to publish. Instead, like many of his colleagues, he sought innovative ways to steer his works into the hands of readers he considered capable of mastering the content.

One technique that scientific authors used to help them control readership involved inserting a preface that carefully described the ideal reader for a given book, an approach I examine in chapter 2. Though couched in the rhetoric of many dedicatory epistles of the period, the "Preface to the Reader" or "Advertisement to the Reader" in a scientific book often sheds light on the author's expectations for a work's reception. Some demanded disciplinary expertise or a certain skill level in the field, while others, particularly in the fields of medicine, hoped to reach a broad audience of lay readers. But for every preface that rhetorically opened a work up to the widest possible readership, inviting in all who were interested in the topic, there was another that explicitly rejected such accessibility and railed against the dangers of ill-equipped readers. In the preface to his 1609 *Astronomia Nova*, Kepler lamented that "there are very few suitably prepared readers these days: the rest generally reject such works [as this]. How many mathematicians are there who put up with the trouble of working through the *Conics* of Apollonius?"[22] More pointedly, William Gilbert, writing on magnetism, questioned the point of publishing at all when the work would only be "damned and torn to pieces by the maledictions of those who are either already sworn to the opinions of other men, or are foolish corruptors of good arts, learned idiots, grammatists, sophists, wranglers, and perverse little folk."[23]

These prefaces are performances, to be sure, but they also alert us to the biases authors were most concerned about, and they gesture at the ways they intended to neutralize these tendencies. Some excellent scholarship exists on the reading of Renaissance dedications and prefaces, yet the prefaces of scientific and medical texts remain largely unconsidered. This is a surprising lacuna given the importance of these disciplines. Natural philosophy and its ancillary disciplines influenced early modern society in broad and penetrating ways, cosmologically, religiously, politically, and socially. The prefaces to such works reveal a great deal about how authors thought about their own ideas, about the power of the individuals who would receive them, and about the technology that would transmit them, the printing press.

Prefaces acted as a kind of vestibule where authors could describe who should take up their work, but a more effective approach was for an author to direct their work straight into the hands of their desired readers. Chapter 3 explores the efforts of scientific authors to control the circulation of their published works by distributing free copies to specific readers. This had the dual effect of allowing authors to single out the best audience for their work and to ensure an educated reception of the book before copies were made available for general sale. Christiaan Huygens sent a copy of his *Treatise on Light* to Isaac Newton and Fatio de Duillier at Cambridge, and in the accompanying letter to Duillier he wrote, "I wish above all to know how it seems to you, Sir, who are the most competent judge I know of in these matters."[24] The natural philosopher Gottfried Wilhelm Leibniz also received a copy of Huygens' book, and in a telling example of how Huygens' distribution efforts created a network among recipients, Leibniz wrote to Newton:

> I do not doubt that you have weighed what Christiaan Huygens, that other supreme mathematician, has remarked, in the appendix to his book, about the cause of light and gravity. I would like your opinion in reply for it is by the friendly collaboration of you eminent specialists in this field that the truth can best be unearthed. Now as you also have thrown most light on precisely the science of dioptrics by explaining unexpected phenomena of colors, I would like your opinion about Huygens' explanation of light, assuredly a most brilliant one since the law of sines works out so happily.[25]

Leibniz's reference to "friendly collaboration" seems ironic, given his subsequent vigorous debate with Newton over the invention of calculus, but his comments reflect the kind of cooperative effort that scientists valued. When Huygens sent his book to men like Newton and Leibniz, he linked groups of preferred readers who could challenge, contribute to, or affirm his ideas. What resulted was a complex web of scholarly relationships, built through and around the printed book. Time and again the correspondence of these recipients demonstrates how effective this strategy was.

Rarely, an author could afford to print their own materials, and in these cases controlling circulation of copies became much easier. They still had to distribute their work, but they could avail themselves of such gatherings as the Frankfurt Book Fair or send multiple copies to trusted correspondents who would ensure the works reached appropriate and interested readers. A personal press also freed authors from some of the tedious business that accompanied publication, including negotiating with, and waiting on, commercial printers. Chapter 4 looks at the phenomenon of do-it-yourself printing and engraving by authors in the sciences. While the press constructed in fifteenth-century Germany was both elegant and useful, it was built with large print runs in mind.[26] To obtain more control over the printing process, scientists like Johannes Müller, the mathematician we opened with, as well as Tycho Brahe, Christopher Wren, Johannes Hevelius, and Christiaan Huygens all engineered devices or developed techniques for "small-batch" printing themselves. Often, their efforts dovetailed with their interest in representing information visually. They sought new ways to tether text to image and image to reality. Generally, historians who have considered such innovations have attributed them to a gentleman-hobbyist's mentality, something along the lines of lens grinding or building chronometers, but a closer look at their efforts reveals the connections between impression technologies and scientists' desire to maximize the potential of print. Private presses, and the invention of new techniques for printing, circumvented traditional publishing streams, but they also highlight the way scientists viewed the instrument of the press itself—and the potential for that instrument to shape scientific knowledge.

Finally, for some scientific authors even a personal press was neither practical nor sufficient. They were so disinclined to publish that the prospect of holding back their notebooks and concealing their ideas seemed not just reasonable but preferable. At least they thought so until they encountered a colleague who could help them navigate the complex world of print culture. Those who responded to this need found themselves in a role that lacked a proper name, but which we can understand as editorial. These editors, whose efforts are examined in chapter 5, were often scientists themselves and acted as assistants and amanuenses. They did the work of correction at the same time they negotiated print contracts and arranged for book sales. On top of all that, they often provided more than a measure of moral support, propping up authors who were more inclined to shelve their work than see it into print.

Despite their vital role as mediators between printers and authors, early modern editors have been largely erased in the histories. Adrian Johns, among others, has explored the complexities of the early modern print shop with a focus on the printers themselves. His nuanced examination is foundational to understanding why editors became so important. This study, while not directly considering the perspective of printers, advances Johns' work by tracing the arc from printers to editors, a line that necessarily passes through the authors. Without the help of editors, William Harvey's work on the circulation of the blood would likely never have been published, and Newton's groundbreaking work on universal gravitation would have remained a stack of scribbled notes on his desk. There are other examples of scientists who, wary of publishing or simply uninterested in its encumbrances, enlisted an intermediary to help them negotiate the world of print. Chapter 6 considers the importance of those individuals who acted as editors, or "midwives," to scientific authorship. Their responsibilities varied, but collectively they redefined the role of the editor in the seventeenth century, and their actions proved instrumental in the publication of major scientific works.

That print culture had a clear impact on the sciences is inarguable. Its effects were felt immediately and reverberated for centuries, like the very scientific theories it impressed onto paper. But natural philosophers and mathematicians, astronomers and mechanists, came to print with differing attitudes and ambitions, and those informed the ways they used the technology. What it meant to share a new mathematical or astronomical idea effectively was hardly codified, and the question was far more complex than whether to print or not. Rather, it was how to best engage with print culture to achieve certain ends. This book is as much a history of communication in the early modern scientific community as it is a history of attitudes toward print technology. That the printing press was integral to the early modern scientific community is eminently clear. But like the microscope or telescope, whose very development would be announced in print, the press itself was an instrument that had to be "learned" by its users, and the epistemological status of printed materials was created; it did not exist prior.[27] Metal type did not make "matters of fact." Scientific authors had to reimagine their works as products of the press in a way their early Renaissance predecessors had not. This did not affect their activities much, either as observers, as mathematicians, or experimenters, but it profoundly affected their role as reporters of their findings.

In response to such concerns, they developed techniques to ensure that the press would best serve them. For some, it meant capitalizing on the replicative power of the press. They embraced a broad audience, a public eager to learn more about the earth, nature, or their own bodies. For others the press created a sense of vulnerability, even irritation, which authors negotiated in different ways. As subsequent chapters will show, many scientific authors attempted to keep the lay audience at arm's length. Thus, a gulf opened between scientists, who shared ideas and language and epistemology, and the lay public who attempted to access scientific texts. This chasm was not solely the result of the increased complexity of scientific ideas. It was also the outcome of a concerted effort on the part of scientists to shield themselves from readers who, in their opinion, were ill prepared to engage with their ideas and therefore posed a liability to their careful consideration and acceptance of those ideas. The gap between a scientific culture and the broader public was not an accident of conceptual evolution; it was a circumstance fostered by the scientists themselves. Above the door of Plato's Academy was the philosopher's appeal: "Let no one ignorant of geometry come under my roof." Early modern natural philosophers and mathematicians, confronted with the printing press, reimagined this maxim, inviting a limited few into their works and hoping to keep the rest out. Their efforts are a reminder of the dynamic ways information about science, math, and medicine entered the broader academic and public conversation.

Chapter One

Authorial Attitudes toward Print

> You seem, my book, to be looking wistfully toward the booksellers'
> quarters, in order, forsooth, that you may go on sale, neatly polished . . .
> [Y]ou grieve at being shown to few, and praise a life in public, though I did
> not rear you thus. Off with you, down to where you itch to go. When you
> are once let out, there will be no coming back.
> —HORACE, EPISTLE 1.20 (20 BCE)

In the preface to his work *De Magnete*, published in 1600, the English natural philosopher William Gilbert wondered why he should bother to publish only to be "damned and torn to pieces."[1] Gilbert was not alone in asking this question. Natural philosophers, mathematicians, and astronomers, confronted with the reality that their audiences were no longer limited to fellow academics, questioned the utility of publishing. But it was more than a concern over lay readers, who might be unfit to assess their work on technical grounds. Scholars increasingly understood that their printed works would reach well beyond a closed circle of peers, and while that proved advantageous in some contexts, it was also a potential liability. John Locke distilled the situation: "Every thing does not hit alike upon every Man's Imagination . . . [I]t must be dressed another way, if you will have it go down with some, even of strong Constitutions."[2]

Scientific authors did dress things in different ways for different readers, sometimes targeting peers of equal intellectual standing, other times widening their aim to a more general audience. The aim of this chapter is to gain a better understanding of authorial attitudes toward print as revealed in private correspondence and published works. The epistolary exchanges between natural philosophers, physicians, and others provide insight into how they arrived at the decision to publish and how they viewed the printing process itself, from finding a publisher to arranging the details of printed pages to establishing fees. Their correspondence also illuminates strategies of dissemination and authorial attitudes about their audience.

If early modern scientists formed opinions on the challenges inherent in printing their work, they simultaneously cultivated strategies for navigating those hurdles. Few scholars could afford the most drastic posture: a wholesale rejection of print and a refusal to publish. But many practitioners of science found it useful to maintain publication pathways that did not involve print. Specifically, they participated in a robust exchange of manuscripts that preserved an insulated corridor of trade for ideas. The Republic of Letters was built on tens of thousands of letters, exchanged across the Continent and England.[3] Print was not an immediate threat to this epistolary network, but when audiences for printed works expanded, I argue that letters gained a new importance, particularly in the sciences. They acted as a buffer between scholars in a specific field and a broader readership that sat outside of authorial control.

Finally, I consider specific techniques scientific authors used to publish ideas without revealing them. These included distributing sealed letters, often containing handwritten solutions to problems, to trusted confidants, along with requests not to open the contents until a specified time. Scientists also availed themselves of codes and anagrams, simultaneously revealing and concealing their theories. Tethered to ideas of priority and credit, these anagrammatic publications reflected creative responses to the pressures of print culture.

The Imprint of Authority

Before exploring the ways authors manipulated print to serve their needs—or maneuvered around it altogether—it is critical to understand both the meaning and operations of print culture in this period. Many historians have diligently examined both the technical aspects and the social ramifications of print. The former include highly detailed treatments from bibliographers and historians of technology. One of the best came from Philip Gaskell. His *New Introduction to Bibliography* still holds up as a definitive study of book production, covering everything from type creation to presswork, illustration to binding.[4] Works dealing with the social and cultural impact of print are likewise numerous. Lucien Febvre and Henri-Jean Martin's *The Coming of the Book* remains an important study of the interaction between the technology and print, the economic context of its use, and the social consequences of its spread.[5] Works such as these are foundational to an understanding of the practical challenges printers faced and the incredible rate at which they overcame them.

And overcome them they did, as even conservative estimates show. From the first printed book of 1455, the method of artificial writing spread rapidly. By 1500 there were presses established in at least 280 cities that, collectively, printed millions of books.[6] Today, more than thirty thousand extant editions of incunables have been cataloged by the British Library, and the census continues.[7] Such tremendous production naturally gave rise to new material, legal, and intellectual issues, and on this the scholarship is also rich. As we saw in the previous chapter, Elizabeth Eisenstein laid the groundwork for historians of science to think critically about the way that printing affected the growth of natural knowledge in the early modern period. She points to a vast network of scholars who relied on a similarly expansive network of published books. Without the press, Eisenstein argued, science would have continued to inch forward but in small, incremental ways.[8]

The importance of Eisenstein's original two-volume study cannot be overstated, both because of the way she tethered the history of science to book history and because of the scholarship she inspired. Historians turned back to primary sources in early modern science, probing aspects of print culture with nuance and originality. One such scholar was Henri-Jean Martin, whose examination of French printing in the seventeenth century shed light on the topics that writers and readers were interested in and on how authors attempted to reach a new and changing audience for printed works.[9] Martin looked specifically at scientific works—the medical literature of the faculty at the University of Paris, the chymical treatises of Paracelsians, and the mathematical works of men like René Descartes and François Viète—considering both the content and form of their publications. His study reminds us of the important changes these authors confronted. First, the number of scientific works in print was increasing, as his quantitative analysis demonstrates. Likewise, the audience for scientific works was rapidly expanding, though it did so at the expense of technical content. Viète, for example, published affordable math textbooks aimed at the general reader, in what Martin calls an effort at popularizing.[10] Scientific authors also thought more strategically about how their work was printed. Some printed treatises in smaller-format octavos and duodecimos, making them economical and convenient to carry around. Others printed their ideas as broadsides or pamphlets, both of which satisfied a growing public interest in science at a more affordable price. Blaise Pascal's *Essay on Conics*, for example, was printed in 1640 as a broadside and posted around Paris, and astronomical maps—of the lunar surface, for example—were also commonly printed as broadsides, posted publicly, and shared by their author with friends.[11]

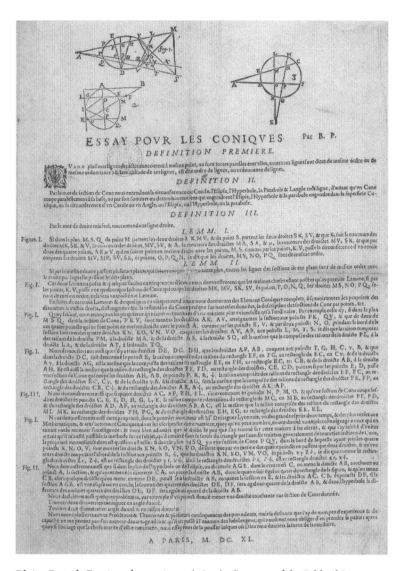

Blaise Pascal, *Essai sur les coniques* (1640). *Courtesy of the Bibliothèque nationale de France.*

In similar fashion, the Elzevirs published a broadside for Descartes entitled *Comments on a Certain Broadsheet*, and in a more pragmatic vein, *On the handling of the large sundial in the Temple of St. Petronius, Bologna* was a didactic broadsheet produced in 1576 to instruct those hoping to use the instrument.[12]

Clearly, scholars in various scientific fields found efficient and effective means of using print technology.

Print Culture and Its Discontents

Martin's study of French scientific printing was complemented by scholarship on English printing practices, particularly where the sciences were concerned. Historian Adrian Johns made one of the most substantial contributions with *The Nature of the Book*. Published nearly twenty years after Eisenstein's monograph, Johns' work reexamined the way printed materials were produced and the processes by which authority was assigned to them. Johns' arguments about epistemology and print are compelling and certainly helped to bolster historians' understanding of scientific knowledge as socially constructed. He also illuminates the early modern concern over plagiarism—especially problematic for new scientific theories—at a time when authors could register neither ownership nor grievances. To put one's manuscript in the hands of a printer was to gamble with the premature spread of those ideas, if not their outright theft. But it is Johns' emphasis on the instability of texts that is most germane to this study. What *The Nature of the Book* demonstrates above all else is the complex, imperfect, and often chaotic process of publishing in the seventeenth century. The London print shop he depicts was a hive of composition, impression, and revision. It was also the site of misspellings, mistakes, and piracy. As Johns puts it, "Printed texts were not intrinsically trustworthy. When they were in fact trusted, it was only as a result of hard work."[13] It fell to printers to secure for themselves, and the works they produced, both credibility and authority. And scientific authors? Johns contends that "faced with using the press to create and sustain knowledge . . . [they] found themselves confronting a culture characterized by nothing so much as indeterminacy."[14]

Johns' work is an important reminder of the many perils that surrounded printing. From the moment a manuscript was deposited in a print shop to the moment when a reader scanned the pages and formed opinions on the treatise, the authorial intention for a publication was at risk. Correspondence from the period confirms this, as natural philosophers, mathematicians and others weighed their reluctance to print against the need to get their ideas into the public domain. Mathematician and astronomer Regiomontanus voiced concern in 1494 about "infecting posterity with erroneous copies of books" because he knew well that "the admirable art of printing" could just as easily multiply erroneous works as publish properly corrected editions.[15] Letters between contemporary authors and their colleagues reveal their varying attitudes toward print, as some worried about making their work public while others were eager to embrace the technology. Once they decided to publish, they discussed everything from the choice of a printer to a book's desired appearance to the target audience for a work. They considered who was likely to read a work, who they *wanted* to read it, and how they might improve their odds of gaining a favorable reception. And of course, they worried about issues of priority. Publishing was a clear—though hardly guaranteed—method of establishing priority for a discovery or theory, but submission of a manuscript to a printing shop was inherently risky. Printers, journal editors, or clerks often found the opportunity to leak new ideas tempting and lucrative, and authors had little recourse.[16] The act of securing credit for a scientific idea was sometimes the very thing that guaranteed credit would slip away.

Johns was not the only scholar drawing attention to the sources of anxiety that surrounded printing, though his picture of seventeenth-century English printing was especially detailed. Bibliographers like D. F. McKenzie took a more holistic approach to understanding how authors felt about print, highlighting a level of authorial anxiety that pervaded the early modern period. If there was a loss of intimacy in shifting from the spoken to the written word, McKenzie argues, then an even greater gap opened up with printing, as ideas passed from hand to paper to press. Poets and scholars might be comfortable sharing manuscripts with a closed circle of colleagues but not with seeing those works broadcast widely in print. McKenzie points to men like John Donne, who refused to print: "What presses give birth to with sodden pangs is acceptable, but manuscripts are more venerated."[17] English jurist Matthew Hale argued the same, expressing his desire to keep his papers out of print because they "are not fit for every man's view, nor is every man capable of making use of them."[18] McKenzie identified "moments of anxiety and of hesitant adjustment" on the part of authors confronting print culture, and these moments merit further examination, particularly when they come from scientific authors.[19]

A Catalog of Concern

The above scholarship provides context for a deeper probe into the ways scientific authors viewed the printing press. By developing a "taxonomy of attitudes" held by early modern authors toward print culture, we can begin to understand the motivations of Martin Lister, when he agreed to share his work on plants with Oldenburg but stipulated that "nothing of this nature from me may be made publicke by ye presse for quiet sake."[20] We can make sense of Isaac Newton's remark that he considered a work to be "too straight and narrow for publick view."[21] In their correspondence and diaries we can locate many of the usual concerns: hesitation related to censorship and legal ramifications, fears that their work would disappear in a sea of printed material, frustration with logistical issues related to the print shop, and a real concern about piracy in the form of abridgments or epitomes that could deny them credit in its myriad forms. As Adrian Johns succinctly puts it, "Would-be scientific authors had to confront a culture of print that they believed to be riven by anti-authorial conduct."[22] I will examine these concerns in some detail below, but such a list is hardly exhaustive. Scientific authors also worried about the preparedness of their readers and the hazards of an unlearned audience. They expressed unease with the publicity that often followed printing, noting its deleterious effects on a contemplative lifestyle. While some authors thrived on using their printed works to catalyze public debate, many complained to colleagues about the demands such disputes made on their time. Scientists also wondered what purpose and whose interests print served. Michael Hunter, in his work on Robert Boyle's publications, describes the balance scientists struggled to find between operating cooperatively, in the spirit of Bacon, and pursuing their own individual agendas. Bacon had called on natural philosophers to work together, openly and publicly, so that the accrual of scientific knowledge was not slowed by selfishness or secrecy. At the same time, scientists needed to take individual risks if they hoped to secure credit for their successes. Thus the tension that Boyle and many of his peers felt. There was also the issue of notoriety, a curse for those few who were simply shy and preferred to keep their names—if not their ideas—out of circulation. By probing these attitudes and teasing out the nuanced ways scientists approach print culture we improve our understanding of not only how science worked in this period but how practitioners made the instrument of the press work for them.

CONCERN #1: CENSORSHIP

There was no better reason to avoid the printing press than censorship. As any early modern author knew, having one's work appear on the church's *Index of Prohibited Books* was hardly desirable. A formal list of banned books appeared first in Paris in 1544, and similar indices were subsequently developed in Italian cities under the aegis of the Roman church.[23] Not only were individual works banned, but an author's entire corpus could be listed, including any future treatises. Printers, publishers, and booksellers could also find themselves banned, having been implicated through sales catalogs or even the identification of a font of type they owned.[24] Many of the prohibited books were obviously religious in nature, as fault lines emerged between Protestant and Catholic presses, but the sciences were hardly immune to censorship. Copernicus' *On the Revolutions of Heavenly Spheres* was banned in 1616 pending revision of certain claims, but it was Galileo's condemnation in 1633 that reverberated across Europe, unsettling other astronomers and natural philosophers. René Descartes learned of it while residing in the Low Countries and was shocked enough to withdraw his own publication plans for *The World*, telling his colleague Marin Mersenne, "I preferred to suppress [*The World*] rather than to publish it in a mutilated form. *I have never had an inclination to produce books* . . . [I]f my views are no more certain and cannot be approved of without controversy, I have no desire ever to publish them."[25] He insisted that, although he held positions similar to Galileo's, "I would not for anything in the world wish to sustain them against the authority of the Church."[26] Descartes was so shaken by Galileo's arrest that he threatened to burn his papers. Stephen Guakroger points out that even if Descartes had published in the Netherlands, where printers enjoyed greater freedom, his work would have been censored in Catholic France, just as Galileo's had been in Italy.[27] For Descartes, it was not enough to simply avoid censors; he wanted to publish work that would be acceptable to them. In closing his letter to Mersenne, he reminds his friend of something the Roman poet Horace said, "Keep back your work for nine years," and he jokes that he has only been working on his treatise for three years, so Mersenne should be patient. But the reference was not entirely in jest. The full quote from Horace says that an author should "put your parchment in the closet and keep it back till the ninth year. What you have not published you can destroy; the word once sent forth can never come back."[28] Printing made Horace's admonition that much more acute, as Descartes understood. Numerous natural philosophers responded with similar concerns, as their mechanistic and materialist philosophies of nature put them at odds with the church. The first part of Pierre Gassendi's *Exercitationes Paradoxicae*, published in 1624, challenged Aristotelian ideas about nature and matter, but despite Gassendi's efforts to abide by church doctrine in his natural philosophy, critics promptly attacked him. Gassendi told a colleague that the book had nearly "provoked a tragedy" among Aristotelians, and he decided to discontinue publication of the work thereafter.[29] Authors whose work brushed up against areas of magic or mysticism—alchemy being only one example—were likewise leery of the attention print would draw.

Concern #2: Audience Preparedness

Not all readers possessed the skills to judge a scientific work. This posed a challenge to authors who both needed and sought constructive feedback to advance their ideas. Christiaan Huygens, for example, was relieved to find that an English colleague could assist with some of his ideas in physics: "I was very glad that Lord Brouncker has taken the trouble to examine my demonstrations of the Laws of Percussion and his approval of them makes me the more obliged, because I find there are few who are capable of judging them."[30] Likewise, when astronomer Johannes Hevelius published his *Cometographia* in 1668, he immediately dispatched copies to Royal Society astronomers in England, along with an eager request that they share their judgment on the work. And Leibniz tells Oldenburg in a letter, "What I particularly beg is . . . to ascertain the opinions, comments, remarks, addenda, corrigenda, and criticisms of those distinguished men upon that paper as each may think it worthwhile."[31] He goes on to list ten specific men—Ward, Wren, Wallis, Lower, and others—whose opinion he would most like to see. Through their efforts, he said, his "rather crude theory" might be refined and its demonstration strengthened.

At the same time, authors expressed apprehension about readers who lacked the technical skills to understand a work but who nevertheless were keen to engage in public discourse on scientific topics. The practice of epitomizing, or condensing books down to a readable summary, was a clear response to this dynamic. Works were stripped of their complexity and distilled to an accessible length, that they might "wittily," "pithily," and "plainely" present information to the reader.[32] Epitomes in law, philosophy, or anatomy were also generally cheaper than full-length works. John Wallis, however, worried that such abridgments might "endanger the loss of the author himself" because they saturated the market for a work before the author had a chance to see a full treatment published.[33]

Both the existence of these epitomes and authors' unease with them speak to the notable growth in the audience that sought out scientific works. Whether an individual's interest was cultivated in formalized scientific societies, at universities, or amid the more casual environment of coffeehouses and salons, it is evident that by the seventeenth century the demand for scientific works had risen. Moreover, the ideas and theories articulated in these works were up for debate. In the spirit of the "new science" originally outlined by Francis Bacon, natural knowledge should be the product of an open process, both experimental and discursive, one that rejected secrecy in favor of transparent consideration of evidence.[34] He called for a shift from esoteric to public knowledge. "Public," however, did not imply a work should be universally accessible but rather that it should target a limited audience that included the monarch, natural philosophers, and learned gentlemen.[35] This conception of knowledge-making dovetails with the notion of the "public sphere" that arguably took shape in the early modern period. As originally described by Jürgen Habermas, the emergence of the new public discourse was contingent on increased literacy and the rise of shared spaces like salons, coffee shops, and organized academic gatherings.[36] Habermas focused on the economic context that gave rise to public exchanges, but he gave less attention to the intellectual, social and technical conditions that facilitated them. He viewed science, for example, as a field external to public opinion, citing Hegel's position that science "is not a matter of clever turns of phrase, allusiveness, half-utterances and semi-reticences, but consists in the unambiguous, determinate, and open expression of their meaning and purport."[37] For Hegel, as for Habermas, one could not debate empirically demonstrable facts. Scholars have since advanced Habermas' study in important ways, particularly by attending to the role of print in facilitating the creation not of a singular public sphere but of several spheres, each of which fostered the exchange of ideas.[38] Recent scholarship reconsiders the way modes of communication, both in manuscript and print, influenced the formation of various audiences for science.[39] These audiences were more heterogeneous than those that earlier generations of natural philosophers had encountered, a fact with tremendous consequences for their publication efforts.

Early modern scientists generally welcomed criticism if it was grounded in the methods of the field, though one's gender and class certainly ascribed credibility in different measures. Robert Boyle, one of the founding members of the Royal Society, believed in man's ability to penetrate the secrets of nature through public experimentation. He also advocated for vigorous debate about natural phenomena, so long as that discussion took place among trustworthy men willing to come to a consensus about observations.[40] But Boyle and his peers thought differently when it came to addressing critiques or attacks from a reader who did not grasp the technical details of a work. Scientific authors found such scrums tedious and a waste of time, but they could be avoided. Galileo Galilei provides one of the best examples of a scientific author who reflected on the readership of his work and consciously attempted to steer it in a particular direction. In his 1615 letter to the Grand Duchess Christina of Tuscany, Galileo wrote candidly about unlearned readers who considered themselves educated enough to proffer their opinions on scientific ideas in print, or who believed that they were equipped to read and react to the writings of scientists and philosophers. He refers to those who "are clearly seen to lack that understanding which is necessary in order to comprehend, let alone discuss, the demonstrations by which such conclusions are supported in the subtler sciences."[41] Galileo goes on to explain how scripture is often used as a tool to dispute scientific evidence, with biblical verses sprinkled throughout the argument to give it an appearance of legitimacy, despite the fact that the attacker is familiar with neither the Bible nor the science. This merely yields conclusions that "are repugnant to manifest reason and sense." Compounding this problem, he argues, is the fact that those unfamiliar with scripture and natural philosophy far outnumber those who truly understand both. This creates an uneven field, wherein "the smaller number of understanding men could not dam up the furious torrent of such people, who would gain the majority of followers simply because it is more pleasant to gain a reputation for wisdom without effort or study than to consume oneself tirelessly in the most laborious disciplines."[42]

Through his letter to Christina, Galileo was clearly making a broader argument to the ecclesiastical hierarchy, which cast a long shadow over his work and his potential publications. Scholars also note that the selection of Christina as his reader gave Galileo "the opportunity to address the lay public in general, a kind of secondary audience that contained the politically powerful, as well as mathematicians and philosophers like himself."[43] And for this audience he hoped to clearly delineate between the trained philosophers and the enemies of learning who, he wrote, "cast against me imputations of crimes, which must be and are more abhorrent to me than death itself."[44]

Conversely, there were women in the sciences who hesitated to publish not because they thought readers would not be prepared for the work but because they would reject the findings out of hand if they did not come from a man. Camilla Erculiani, an apothecary and natural philosopher, authored an alchemical work in the late sixteenth century wherein she acknowledged that many readers would assume alchemy was "a subject that does not belong to women (according to the custom of our age)."[45] She rejected this position but noted that "the work of caring for my children, the burden of running my household, my obedience to my husband, and my fragile health—none of these weighs on my decision to publish so much as the knowledge that many malicious minds will condemn my efforts, and writings, and consider them frivolous and worthless, just as they consider women of our age to be such."[46]

Concern #3: Too Many Books!

An unlearned public was clearly a threat to those publishing new ideas in the sciences, but the inverse problem existed as well: books were also being published at increasing rates by authors whose skills in mathematics, astronomy, or natural philosophy varied. The problem, of course, was not confined to the sciences. Renaissance humanists like Erasmus believed that an avalanche of new books, which were "foolish, ignorant, malignant, libelous, mad, impious and subversive," threatened scholarly attention on the classics, which he and others held in great esteem.[47] His worry was not without reason. The sheer number of books coming off European presses was astounding. Between 1580 and 1712, the Dutch printing house Elsevier, whose catalog included numerous scientific works, ushered sixteen thousand editions into print.[48] In 1651 Oxford scholar Robert Burton bemoaned the avalanche of books coming off English presses. "Many are possessed with the incurable itch for writing," a state that yields countless numbers of books. "New books every day, pamphlets, curantoes, stories, whole catalogues of volumes of all sorts, new paradoxes, opinions, schisms, heresies, controversies in philosophy, religion, etc . . . 'presses be oppressed.'"[49] Burton raises his hand, he says, like a felon pleading guilty to the crime of adding to the crush of books: "I am content to be pressed with the rest." Though poetic, Burton was hardly hyperbolic. The historian Ann Blair, in her study of information overload in the early modern period, has shown how scholars devised new systems of information management to organize what they perceived to be an overwhelming increase in available knowledge. Even in natural philosophy, Blair writes, the sheer number of works was enough to overwhelm readers, and new techniques to deal with the scope of printed material were proffered.[50] It is no wonder, then, that a natural philosopher like William Gilbert was so reticent about putting his work into print. In *De Magnete* (1600), he excoriates the state of publishing, asking himself, "Why . . . in so vast an Ocean of Books by which the minds of studious men are troubled and fatigued, through which very foolish productions the world and unreasoning men are intoxicated, and puffed up, rave and create literary broils, and while professing to be philosophers, physicians, mathematicians and astrologers, neglect and despise men of learning: why should I, I say, add aught further to this so-perturbed republick of letters."[51]

Gilbert's concerns were widely shared. Take the case of mathematician John Pell, who formulated methods to deal with the crush of mathematical books being printed. In a letter printed as a broadside in 1638, Pell writes about the difficulty with achieving an "advancement of mathematickes."[52] His hope was that the creation of a complete inventory of mathematical works would ensure that materials were readily available to those who wanted to study them. To accomplish this, he outlines specific steps: (1) creating a synopsis of the mathematical writings "either *extant* in print, or *accessible* manuscripts in public libraries"; (2) a complete and chronological listing of all mathematicians of note, "with the year when any of their works were first printed"; and (3) "a catalog of the writings themselves, in the order of years in which they were *printed* in any language . . . adding the volume . . . not only what fold [40. 80. &c.] but also the number of leaves, that we may estimate the bulk of the book."[53] Pell's letter acknowledges that manuscripts housed in public libraries are useful, but it is clear that print has become the standard medium. In itself, this is unsurprising, though bibliographers will appreciate Pell's attention to the book's format—folio, quarto, octavo—as a partial measure of a book's density. What is more interesting is the way Pell suggests that students of mathematics deal with the number of books available, which by the early seventeenth century was formidable. He proposes that a council be established to help students determine not only which books are the best for a given area of math but "in what *order*, and *how* to read them, what to observe, what to beware of in some mathematicasters [*sic*], how to *proceed*." This council would also assist in the development of a catalog that effectively ranked books in order of importance so that "men might be informed, in that multitude of books, with which the world is now pestered, what the *names* are of those books that tend to this study only."

Pell also called for the erection of public libraries that would contain the entire body of mathematical writings, including works printed in other coun-

tries. But, he conceded, such an aggregation of materials might not be sufficient to help the poor student. His letter considers the possibility of composing what he calls "An Instruction," which would show any mathematician how
to do problems without the help of a library so that "we be no longer tyed to
books." What this instruction manual would contain is not clear, but Pell is
convinced that proper attention to method could yield a single work, a "Pandect," which would give a student precisely what they needed to learn in the
event that a library was not available. The motivation for creating such a work
comes from Pell's sense that print has simply flooded the intellectual landscape
with materials. He claims his Pandect would "spare after-students much *labor
and time* that is now spent in *seeking* out books, and *disorderly reading* them,
and *struggling* with their cloudy expressions, unapt representations, different
methods, confusions, tautologies, impertinencies, falsehoods by paralogisms
and pseudographemes, uncertainties because of insufficient demonstrations,
&c. besides much cost also, now thrown away upon the multitude of books,
the greater part whereof they had perhaps been better never to have seen."[54]

Pell's letter reflects a common sentiment at the time: not that books and
print were bad but that print allowed for the number of books to increase to
an untenable point, a point that made nurturing new mathematicians difficult. Without careful guidance about what they read and instruction on how
to process that information, there was little hope of advancing the field. His
is not an expression of personal reluctance to print but a broader indictment
of the circumstances print had fostered. As his mathematical colleague Leibniz puts it in a letter to Louis XIV, "the horrible mass of books that keeps on
growing" robs authorship of any honor and threatens to bury readers in dis-
order.[55] Like Pell, Leibniz called for the creation of a "quintessence of the best
books" to be stored in academies.[56]

The number of books, then, was clearly a problem. But it was not simply
an issue of charlatanism or pseudoscience joining the slipstream of publication,
since the very definition of what constituted science and what did not was
being negotiated. What did and did not constitute science was hardly a shared
understanding or a fixed entity. Consider the publications listed in the 1673
catalog titled, *All the Books Printed in England Since the Dreadful Fire of London
in 1666*. Published by bookseller Robert Clavell, this catalog covered the years
1666–1672. In the section on physick, one finds *The Diseases of Women with
Child*, a French work translated by English physician Hugh Chamberlain. In

the same section the catalog lists *Secrets Revealed: Or, An Entrance to the Shut-palace of the King*, a work attributed to "a most famous English man who . . . attained to the Philosophers Stone, at his Age of Twenty three years."[57] A modern reader would question the veracity of the latter, but such a judgment was not necessarily made at the time. Likewise, a 1683 catalog offers the same combination of scientific effort and seeming quackery, pairing a *Chymical Secrets, and Rare Experiments in Physick and Philosophy*, which promises to include "philosophical arcanum," alongside *Vates Astrologicus, or England's Astrological Prophet*, which claims to reveal Great Britain's future.[58] Anyone, theoretically, could publish their ideas if they could secure the resources, and with so many works being printed, the growing audience for scientific works could scarcely navigate them all.

In a letter to Samuel Hartlib, Robert Boyle seems positively relieved to have found a "knot of such ingenious and free philosophers" at Oxford who were studying and talking about experimental philosophy. Boyle said such work would contribute more to the advancement of learning "than many of those pretenders that doe more busie the presse & presume to undervalue them."[59] Physician Timothy Clark distilled the issue nicely: too many works were being printed by too many people who did not, in his estimation, have something meaningful to contribute. This created, in Clark's opinion, a precarious situation: "The world of learning totters and is almost overthrown so long as many seekers after the ear of the crowd, serving neither truth nor human welfare, daily obtrude upon the world their absurd ideas, ill understood and perhaps scarcely to be understood. Whence over precocious minds are crazed by incomprehensible notions."[60] A host of "mountebanks, conjurors and peddlars," Clark writes, were brazenly sharing their views in print without any sort of reproach from the learned community. The minds of the uneducated quickly become addled amid so much information, incapable as they were of sifting the valuable knowledge from the bunk. Clarke unequivocally summarizes his attitude toward print: "If only matter sent to the press were submitted to a mature judgment, learned men would not labor under such a great mass of useless books."[61] He even acknowledges that, despite a decade of experimentation in medicine and the encouragement of colleagues, he was reluctant to print his own findings.

Meanwhile, Englishman Barnaby Rich—no scientist but a keen observer of publishing trends—underscored the fact that interesting and intelligent books would fail to sell in such a flooded market without something to seduce readers, either an alluring title or a captivating frontispiece:

> For he that will presume to publish a booke, if hee doth not learne with the Tayler and the Atyremaker, to put it into a new fashion, it will never sell, it will lie still in the Printers hands, and those lines that are now put in print if they conteine any matter of pietie, or that are any whit at all entending to honesty, they doe but pester a Stationers stall, and there are very few or none that will bestow one peny on them, such is the curiosity of Opinion in this age, but especially concerning bookes.[62]

Faced with the dual challenge of a flooded market and a broader, less academic audience, some authors consciously tailored their works for a particular audience. Returning to the exchanges between Mersenne and Descartes, one finds Descartes carefully considering the kind of readers he wanted for his work and those he was likely to get if he failed to shape his texts in certain ways. Writing to Mersenne in October 1629, Descartes asked him not to speak about his current project, a small treatise called *Meteorology* that would be appended to his larger project *Discourse on the Method*. Descartes wanted his authorship of the work to remain unknown, explaining, "I have decided to publish this treatise as a specimen of my philosophy and to hide behind the picture in order to hear what people say about it . . . and I shall try to expound it in such a way that those who understand only Latin will find it a pleasure to read."[63] Descartes clearly sought an educated audience, based on his preference for writing in Latin, but he also claimed to want anonymity.[64] Publishing anonymously was not unheard of in the early modern period, though it has been considered "no more than an ephemeral feature of the text," a temporary screen at best.[65] Thus, it is questionable how anonymous Descartes could have remained, given his reputation and the awareness among intellectuals of his philosophical work. At that point in his writing, however, he was insistent on leaving his name off the treatise. In a subsequent letter he told Mersenne that he would prefer if nobody knew he was working on it, so that he could think and write in peace without dealing with the expectations of others. Descartes was adamant that he did not seek fame; on the contrary, he claimed to fear it. "I think that those who acquire [fame] always lose some degree of liberty and leisure, which are two things I possess so perfectly, and value so highly."[66] He assured Mersenne that no amount of money could persuade him to give up the freedom he enjoyed when working and publishing out of the public eye. More than a year later Descartes was still asking Mersenne not to speak of his project: "If any people should have the idea that I am intending to write something, please disabuse them of the idea . . . and make it clear to them that I have absolutely no such intention."[67] He maintained this posture seven years later, when he handed off the work to a printer.

Two things emerge from Descartes' correspondence, both related to the scientific audience and the author's privacy. On the one hand, Descartes wanted the freedom to write his treatise out of the public eye "to avoid being disturbed and to keep the liberty I have always enjoyed." Aside from promising the manuscript to Mersenne within three years, he preferred to steer clear of deadlines, "so that no expectations may be raised and my work may not fall short of expectation."[68] If he had been writing before print, the pressure to produce would have been significantly reduced. How many people would wait eagerly for the forthcoming manuscript of a natural philosopher? The slow pace of scribal copying made such expectations impossible, but print compressed the timeline. Throughout the correspondence of the early modern period we see scholars anticipating the publication of someone's latest mathematical treatise or astronomical tables. Print runs of hundreds or even thousands of copies, coupled with widespread dissemination of these books across the Continent, dramatically increased the odds of getting one's hands on a scholar's most recent work. Oldenburg, for example, was a conduit for hundreds of scientific works during his career. Sometimes he included a copy of a recently published work in his letter to a correspondent; other times specific works were requested by a correspondent. He was asked by Henri Justel, a French colleague, whether he could procure a list of nine scientific books and the complete run of *Philosophical Transactions* for a Mr. Lantin, councilor of the Parlement of Dijon. Justel asked that Oldenburg purchase the works—which ranged from Francis Bacon's *Miscellany* to Henry Savile's *Mathematical Lectures*—let him know the cost (8s 6d), and then arrange transport.[69] An audience eager for a scholar's latest work created pressure for some authors.

But Descartes also anticipated masking his identity when he finally did publish, a move he hoped would further insulate him from criticism, at least for a time. His reluctance to print had clear links to issues of pressure and privacy. For him, as with other natural philosophers, the perceived obligation of responding to critics weighed heavily once a work was widely circulated. By March 1636 Descartes claimed he was ready to publish a treatise that consisted of his *Discourse on Method, Optics, Meteorology,* and *Geometry.* In his letter to Mersenne that month, he speaks of publication details. The leading scientific publisher in Leiden, where Descartes was living, was the Elsevier firm, but Descartes expresses frustration with them and was shopping for a different printing house. He also describes the way he wanted the book to appear: "I would like to have the whole thing printed in a handsome fount on handsome paper . . . [T]here will be four treatises, all in French."[70] Obviously he had elected to publish in French, not Latin, a decision that had ramifications for his intended audience. Having left out his more controversial treatise on the world, he was more inclined to expand his readership. His proposed title for this publication is a reminder that brevity was not paramount in the seventeenth century: *The Plan of a Universal Science which is capable of raising our Nature to its Highest Degree of Perfections, together with the Optics, the Meteorology and the Geometry, in which the Author, in order to give proof of his universal Science, explains the most abstruse Topics he could choose, and does so in such a way that even persons who have never studied can understand them.* It is a comprehensive name to be sure, but most important is the final line, which signals Descartes' desire to reach a wider audience. In a subsequent letter he explains that he truncated his proof for the existence of God purposefully because he was "afraid that weak minds might avidly embrace the doubts and scruples which I would have had to propound, and afterwards be unable to follow as fully the arguments by which I would have endeavored to remove them."[71] In a letter to another scholar he hoped the treatise was so accessible that even women could comprehend it, a remark meant hyperbolically but also one that belies Descartes' extensive interactions with female readers.[72] His relationship with Princess Elisabeth of Bohemia, for example, seems to have been mutually beneficial. Elisabeth sent him queries on metaphysics and natural philosophy, challenging him in places. Descartes responded at length, fleshing out aspects of his work according to her questions and interpretations. His 1644 *Principles of Philosophy* was dedicated to the princess in recognition of her vital role in his intellectual and philosophical development.[73] Elisabeth was just one of the so-called Cartésiennes, women who took up the study of Descartes' philosophy. Another close connection is seen in Descartes' interaction with Queen Christina of Sweden. The two exchanged correspondence related to Descartes' mathematical and philosophical work, Descartes sent her a manuscript copy of *The Passions of the Soul* (which was also dedicated to Elisabeth), and the two met four or five times in Sweden prior to

Descartes' death in 1650. It is fair to say, then, that Descartes' quip about dumbing his work down for women signaled less about his attitude toward female readers than it did about his desire to compose his argument in an accessible way. Lest Mersenne worry about such distillation, however, Descartes assured him that the full argument, without truncation or simplification, existed and that he might eventually publish, in Latin, the more erudite version for learned readers.

If issues of privacy and audience reception were not enough to challenge authors like Descartes, there was also the sheer tedium of working with printers, who were often in a financially difficult position themselves and trying to balance the bottom line with authors' demands. Those with a commitment to printing scientific works had to negotiate the hurdles of getting the ideas into print economically and then finding a viable market for the works once they were published. Authors were aware of these commercial pressures, though they were not always sympathetic. General discontent with printing houses is expressed by authors from an array of fields. One English cleric resorted to Greek to veil his disdain for the work of printers, calling them stupid and mean (σκαιότη and μικρολογία), and the historian of the Cambridge University Press, David McKitterick, has plainly noted that, "for most books, even those by authors close to the press, accuracy was striven for only intermittently."[74] Descartes struggled over the years with his publishers, most of whom assumed financial responsibility for his works and therefore had to consider such details as licensing and sales. For example, he complained about working with Jan Maire, printer of his *Discourse on Method*, telling Dutch diplomat and poet Constantijn Huygens, "Each day I regret the time which the publications for le Maire have cost me. White hairs are rapidly appearing on my head."[75] To be fair, Maire was trying to execute the printing of Descartes' treatises at the same time the woodcuts for the images were being completed by the Dutch mathematician Frans van Schooten. Typically, one would have the images done beforehand, so Descartes put Maire in a difficult position. Eventually, van Schooten agreed to move into the publisher's house, in part, Maire said, "for fear that he would escape."[76] But providing images was just one of dozens of tasks that had to be completed prior to making actual impressions. Once a manuscript was in a printer's shop, proofs were pulled and then correcting began. This could be tedious, and, as often as not, the author was just as responsible for the slow pace of production as the printer. There were also details related to licensure, as publishers or authors had to secure permission, also known as a *privilege*, to publish a work. In sum, the task of putting one's work out into the world was not for the faint of heart.

Descartes' experience in the Low Countries was hardly unique. His various concerns were echoed by scientists in England. Correspondence from the English mathematician John Collins is illustrative because Collins understood the publishing business so well. Booksellers hesitated to take on scientific works because sales were too often sluggish. What they needed were assurances that enough books would be sold—between eighty and one hundred—to keep them in the black.[77] Collins acted as an intermediary between scientists and the printers and publishers who, with some persuasion, were willing to produce their works. An interaction he had with the Scotsman James Gregory provides an example of the kind of work Collins did, while also illuminating some of the challenges a scientific author faced around publishing. In a letter to Gregory in 1670, Collins relates his conversations with Isaac Newton on such topics as musical progressions, the infinite series, and other mathematical projects. He describes an occasion when he had asked Newton for his thoughts on "the Interest Problem," but Collins says he heard nothing back from Newton and, "observing a wariness in him to impart, or at least an unwillingness to be at the pains of so doing, I desist, and doe not trouble him any more."[78] On the face of it this is simply an anecdote about a natural philosopher who did not have the time at that moment to explore a particular problem. But Newton was known for his reluctance to write up his works, and Collins was not merely asking casually about some mathematics. He was attempting to elicit something more formal from Newton, just as he had coaxed the work of Newton's predecessor, Isaac Barrow, into print. Collins goes on in the letter to express his disappointment about Newton's hesitation, a sentiment others shared. Rob Iliffe, who has examined Newton's publishing efforts at length, suggests that by 1676 Newton "had had his fill of print culture and had long since tired of his critics' attitude to his optical theories."[79] Newton told Oldenburg that he preferred to spend his time in correspondence with other mathematicians and natural philosophers, "for I see a man must either resolve to put out nothing new or to become a slave to defend it."[80] Collins reflects on this reluctance in his letter to Gregory, but immediately after complaining about Newton, he pivots and begins to effusively praise of Gregory's willingness to see his own work into print. "I much rejoyce at your Inclination to publish your lucubrations." To expedite publication he offers Gregory assistance with correcting the printed pages, but he still laments the state of publishing in mathematics. "There is not any printer now in London accustomed to Mathematical work, or indeed fitted with all convenient characters, and those handsome fractions."[81] Collins knew whereof he spoke. He proceeds to enumerate all the different mathematicians who had work in the "queue" of William Godbid, a printer in Little Britain known for producing quality mathematical treatises. With three presses, specialized type, and around five workmen, Godbid could scarcely keep up with the demands of printing the works of astronomers, mathematicians, and geographers, especially given his own financial circumstances at the time, which were not good.[82] His shop was essentially booked until the following spring, a fact that led Collins to suggest that Gregory consider printing his work in Scotland, since it might be faster.

Collins was acting here as a catalyst to publication, a role I will examine further in the chapter on scientific editors. His efforts, however, pale in comparison to those of Henry Oldenburg, secretary of the Royal Society and correspondent extraordinaire. With an abiding belief in the merits of scientific collaboration, he encouraged many natural philosophers, astronomers, mathematicians, and others to make their work public. In April 1662, for example, he received a letter from Dutchman Benedict de Spinoza wherein Spinoza expresses hesitance about publishing on the origins of matter and first causes. He had written a complete treatise on the question and had edited it for publication, but he feared doing so, "lest the theologians of our age take offense and attack me, who utterly detest disputes, with their usual malice."[83] He asks for Oldenburg's advice on the matter. Oldenburg replied a few months later, urging Spinoza to publish his philosophical treatise: "I would by all means advise you not to begrudge to scholars the results you have attained in philosophy and theology through your wisdom and learning; let them be published, however much the would-be theologians may snarl . . . Come then, excellent Sir, and banish all fear of vexing the homunculi of our age; sacrifices have been made to ignorance and absurdity long enough; let us hoist the sails of true knowledge and search more deeply into the recesses of nature, then [sic] men have done hitherto."[84] The following spring, Oldenburg was still pressing the issue, asking Spinoza how the treatise was going and urging him to publish: "I adjure you . . . not to grudge or refuse us your writings on these subjects."[85] This exchange with Spinoza was repeated many times by Oldenburg with other scholars. Among the works he encouraged and facilitated were treatises by Samuel Hartlib, Robert Boyle, Marcello Malpighi, and John Wallis. A kind of literary midwife, he personally helped authors navigate the challenging path to publication, along the way allaying their concerns about the potential audience, the risk to their reputations, the costs of editing, and the potential impacts of censorship. Oldenburg understood the range of concerns regarding putting work into print, and he was a master at helping authors overcome them.

The examples highlighted above point to a variety of reasons authors in the sciences approached printing with reluctance. How they dealt with this varied, but the strategies they employed remind us how resourceful scientists could be when it came to sharing their work. The first and most obvious tool they utilized was also the oldest: the production and exchange of manuscripts. While printing had become the most effective way to distribute hundreds or thousands of copies of one's work, traffic in manuscripts continued with great efficacy. Historians of the book, bibliographers, and historians of science have provided a nuanced picture of the role manuscripts played in the seventeenth century, long after the paradigm of print had been firmly established. Harold Love's now-standard study of scribal publication in the seventeenth century is a valuable reminder of the role that handwritten documents—be they government papers, correspondence between colleagues, or copies of lectures delivered at a university—continued to play, despite the availability and utility of printing presses.[86] Love's work is complemented by more recent scholarship that probes the liabilities of print more deeply. David McKitterick, for one, has shown that printing implied more than sharing a work. It meant an author would release their ideas "in an uncontrolled way with an audience whose extent and nature could never be known, and whose suitability to participate in knowledge was untested by social and intellectual criteria."[87] In this context, the intrinsic value of manuscript exchange increased rapidly, affording the authors a measure of protection. There is also an extensive literature about the role of manuscripts in preserving a culture of secrecy. Robert Boyle, for one, deftly navigated between the public and private by keeping certain diaries and notebooks in manuscript form, while publishing other work he hoped to share with learned readers.[88]

In the preface to his essay *Some Specimens of an Attempt to Make Chymical Experiments*, Boyle describes writing up some of his experiments related to saltpeter, which he then sent to a friend: "And having dispatched that little treatise, it found so favorable a reception among those learned men into whose hands it came, that I was much encouraged to illustrate some more of the doctrines of the corpuscular philosophy."[89] The manuscript was a trial balloon, circulated in advance to ascertain whether a printed book was worth producing and to ensure that it would be well received. Boyle did this with some frequency, engaging in what Michael Hunter describes as "private printing." He would order small print runs, purchase the entire lot, and distribute copies to his chosen audience.[90] A 1688 publication, for example, contained fifty medical recipes drawn from a larger set Boyle compiled. While he gave the copies away for free, he expected recipients—physicians, surgeons, clerics, and ladies who tended to the sick in their households—to share individual recipes and report back on their efficacy.[91] Boyle's distribution of both small, controlled print runs and manuscript works is a good example of using the press strategically, or not at all, depending on what the author required from the work. For Boyle, there were excellent reasons to preserve the tradition of manuscript exchange.

Other natural philosophers and mathematicians adopted similar approaches to sharing their work through correspondence and as more developed manuscripts. Using a network of private and quasi-public channels, scientific authors interacted outside of print culture in an effort to work more efficiently and, ultimately, to use print more effectively. Scholarship in the past decade has paid much closer attention to the workings of the Republic of Letters, from the various protocols of participation to the roles the most active participants played. Many of the newly formed societies, such as the French Academy of Science or the English Royal Society, ran on the fuel of epistolary exchange, which itself was made possible by the development of formalized courier services. Letters to and from members of these scientific organizations were copied and cataloged, becoming part of the official record.[92] They also played an important part in the arbitration of priority disputes, allowing societies to informally adjudicate conflicts based on epistolary records.[93] But the way natural philosophers, astronomers, and other practitioners of science braided together their research, their private writings, and their publishing has not been fully explored. Naturally, these scholars wrote letters and sought feedback, but upon closer inspection it becomes clear that the iterative loop of letter writing and manuscript exchange became an important primer for the *public* publishing process.

Tens of thousands of letters passed through the hands of the period's central correspondents. Before Henry Oldenburg ascended as England's leading correspondent, Samuel Hartlib produced more than four thousand letters involving at least four hundred individuals. He actively encouraged correspondents to publish some of these letters. John Beale, for one, had written Hartlib extensively on matters of husbandry, and with Hartlib's support these missives were "presented to publick view" after all the Latin phrases were translated and the language made accessible.[94] Hartlib's surviving correspondence exceeds three thousand letters and his lifelong cultivation of scientific networks was vital to the expansion of natural knowledge, but he was hardly alone. Across Europe other participants served as key nodal points in the Republic of Letters. They included Nicolas-Claude Fabri de Peiresc (1580–1637), Marin Mersenne (1588–1648), and the German polymath Athanasius Kircher in (1602–1680), each of whom was committed to sharing new findings in an array of scientific fields. Mersenne's network of more than 70 correspondents is well known among those familiar with early modern science, and Kircher maintained correspondence with at least 763 individuals, but Peiresc was by far the most prolific writer. He sat at the center of an epistolary exchange that exceeded ten thousand letters, most of them in the vernacular, to over 500 correspondents. They included Galileo, Pierre Gassendi, and the Parisian scholars Pierre and Jacques Dupuy, brothers famous for creating a space for scholarly study and debate in Paris. And while Peiresc's interests varied widely, from the behavior of tides to the activities of Mediterranean merchants, his extensive network reaffirms the importance of scribal activity in the period. For despite writing a great deal every single day and orchestrating conversations on dozens of topics simultaneously, he did not—as a recent biographer points out—publish his works in printed form.[95] Instead, he shepherded thousands of letters among friends, colleagues, and acquaintances, creating an extensive archive. With Peiresc, as with others in the period, we must "step outside print publication as the mark of a writer."[96]

Letters exchanged through these and other correspondents went through a multistage process that might include duplication, translation, editing, redacting, revision for publication, or some combination thereof.[97] Take, for example, a series of letters to and from John Beale, a natural philosopher who joined the Royal Society and who wrote extensively in two areas: memory and husbandry. In correspondence between Beale and his colleagues Samuel Hartlib, Thomas Goad, and William Brereton, the men discuss *A Treatise on Memory*, by Caleb Morley.[98] Based on contemporary references and catalogs, it seems Morley's work existed only in manuscript form; it was never printed. In one letter, Goad describes receiving "five parchment rolls (heretofore belonging to Caleb Morley) sent up a few days before from Mr. John Beale."[99] He details the size and contents of each of the rolls. The first was eighty-one inches in length and sixteen inches in breadth; the second nearly fifteen feet long and about the same breadth. Goad also describes the way Morley's work is organized in terms of teaching a mnemonic method. In a letter from the same collection Beale writes, "I received, of Mr. W. Brereton, a roll with 3 parchments fastened to it, which I would here copy out." He goes on to synopsize the scrolls, similar to what Goad had done, and then to elaborate much more on Morley's "Mnemonical Probleme." In these exchanges, Beale and his colleagues copy, comment on, and exchange Morley's work, which was never printed. Beale worked similarly with Oldenburg, sending him lengthy essays, such as "Disposalls & Considerations . . . Concerning the Transplanting East Indian Spices, & other their Usefull Vegetables in our Weste Indyes," a dozen or more pages long. These treatments were epitomized by Oldenburg, who then shared those summaries with Royal Society members. Some ended up in print in the *Philosophical Transactions*, others were picked up by publishers and made more widely available. Oldenburg would also copy summaries to send to individuals whom he knew to be interested in the subject matter.[100]

But men like Oldenburg were more than communication conduits. The very ideas that they transmitted through their correspondence were nurtured and developed, their significance amplified. There was, as one historian put it, "an expectation for mathematical ideas to mature through a web of exchanges."[101] It was manuscript that allowed for this, in a way that print could not. The malleability of ideas written on paper was critical to their improvement, and not because the letters were private. Precisely the opposite. Ideas shared through epistolary channels were understood by most to be potentially, if not actually, public, which allowed them to be augmented or challenged. French astronomer Adrian Auzout wrote to Oldenburg, "I see little difference between printing scientific matters contained in letters and showing these same letters to those learned in these matters who can copy them out when they have them on loan."[102] The only exception to this public nature came when an author specifically requested discretion on the part of a letter's recipient. If an author did not want the content of their letters shared, they clearly stipulated as much. Christiaan Huygens sent his correspondent Jean Chapelain information on his theory about Saturn's moons, as well as a description of a new pendulum clock, but Huygens expressly told Chapelain that "it would be better if it was not divulged until one sees all the evidence of the work which I hope to provide soon."[103]

For those who built extensive correspondence networks, a thorough knowledge of context was necessary to advance scientific ideas. Discretion was critical, and a sense of politesse helpful. Professionals like Oldenburg faced daily decisions over which letters, or parts of letters, to share and with whom. The entire epistolary network relied on such judgment. Cultivating discussion around an idea in astronomy or natural philosophy could effectively generate new ideas and challenge existing assumptions. Debate was key tool for advancing knowledge, but if the sharing of letters produced heated priority disputes, this merely impeded progress. The men who orchestrated epistolary exchanges walked a delicate line between provocation and pacification.[104] They catalyzed dialogue but always aimed to prevent unproductive discord, and they could do this because of the malleability of manuscript. It bent to unique circumstances in a way that print could not.

Another illuminating case involves John Wallis and François Dulaurens, two mathematicians who corresponded between Paris and London through Oldenburg. Dulaurens published *Specimina Mathematica* in 1667—"for the most part an unremarkable book"—and sent a copy to Oldenburg, who was expected to circulate the work among English mathematicians for comment.[105] Oldenburg passed it on to Wallis, Savilian Professor of Geometry at Oxford, who responded in March 1668 with his impressions. "The Title, mythinks, promises much more than ye book doth perform. A great part of his first booke, seems to be taken out of Mr Oughtred & my self (though he doth not there so much as name either of us)."[106] Wallis goes on in his letter to dispute Dulaurens' claim that Wallis had issued a problem to all the mathematicians to solve, and he debunks some of Dulaurens' mathematical ideas. "Now when all this is publicke & notorious; I cannot but wonder, with what confidence (or gross negligence,) he should publish, in ye face of all the world, a thing so false." As happened frequently, Oldenburg had parts, though not all, of Wallis' letter published in the *Philosophical Transactions* the next month, and this printed critique eventually made its way back to Paris. Dulaurens wrote Oldenburg promptly to apologize for any misattribution of the mathematical problem in question, and he asked whether it might be possible for other mathematicians—not just Wallis—to provide feedback on his work. Around the same time, Henri Justel, an active French scholar and correspondent, wrote Oldenburg to warn him that Dulaurens was preparing a response to Wallis "which is very salty." He told Oldenburg that "you will see that piece, which he will send you by some friend. I did not want him to send it to you by post."[107]

The back and forth between Wallis and Dulaurens was quickly becoming a case study in publishing strategies, with manuscript letters shared both privately and publicly, then printed, then responded to in further private letters. There was also the gentle suggestion that such sensitive matters should not be shared via the public post but transmitted through networks of friends. Dulaurens' letter was delivered in June 1668 by Secretary of State John Trevor on his return trip from Paris. The response, which Dulaurens had had printed, was formal in tone and structure: *The Response of François Dulaurens to the letter of Dr. Wallis written to the most excellent Oldenburg.*

Henri Justel, for his part, found the whole affair sadly overblown. "It displeases me to see literary men dispute and destroy one another, because that makes them ridiculous," he wrote Oldenburg. And later, "Literary men are too touchy, in my view, and take offense too easily . . . there is no room for hope that they will ever mend their ways. If Mr. Wallis's letter had not been printed [Mr. Dulaurens] would not have cared about it."[108] It was not Wallis' critique per se but its publication that troubled the Frenchman. Thus, it is not surprising that Dulaurens' printed letter elicited yet another response from Wallis, who sent a manuscript entitled "Animadversions on Dulaurens" to Oldenburg. This piece illustrates the delicate balance between the public and private aspects of publications at the time. Dulaurens indicated in his *Response* that he was upset with Wallis for not making public the grounds of his criticism, but Wallis saw it quite differently: "It is for you [Oldenburg] to clear me of this fault. For indeed I brought nothing of either before the public, but laid both before you. At any rate, you asked me for an opinion: I indicated briefly *in a private letter to you what it was,* and why so."[109] Wallis maintains that it was never his intention to address Dulaurens' treatise in its entirety but only to offer up a few examples he considered problematic in the work. That critique was then edited by Oldenburg for publication in the *Philosophical Transactions.* Wallis writes:

> You yourself know best the reason why, when you published part of this (by which perhaps you redressed the injury done to myself), you concealed the rest (for the sake of gratifying him, as I thought) when you might have either published or suppressed the whole. Since he asks for it, I agree to the publication of the whole letter, just as it was written; this is within your power as it was written to you. Thus the learned world may judge whether or not I had just grounds for those criticisms, however succinctly expressed.[110]

This exchange encapsulates much of what transpired in the early modern sciences: trading books, sharing letters with feedback, copying and editing those letters, producing formal manuscripts related to these exchanges, and printing some of the manuscripts for wider circulation. Oldenburg, in the middle of two dueling mathematicians, attempted to appease both, protecting Wallis' reputation while softening the blow to Dulaurens. And while his diplomacy was admirable, in this case it left both parties unsatisfied.

Exchanging letters was the most common way to share ideas outside of print, but to probe a theorem or a theory more deeply authors also produced manuscripts that circulated among a closed group of readers. These manuscripts were frequently accompanied by letters and, depending on the author's instructions, would be limited to a chosen audience or else copied for a wider readership. An illustrative example is seen in the scientific study of the phenomenon of parhelia. Parhelia were observable spots on a solar halo that made it seem as if there were two or more suns in the sky. An account of this effect was recorded by the German astronomer Christoph Scheiner, who witnessed a parhelion outside Rome in 1629. Scheiner sent his account to Cardinal Francesco Barberini, who in turn sent it to the astronomer Nicolas-Claude Fabri de Peiresc. Peiresc ordered multiple manuscript copies of the observation and accompanying commentary, and he sent these copies to several individuals, including Isaac Beeckman and Pierre Gassendi. Gassendi forwarded his copy to the natural philosopher Henricus Reneri, a student of Descartes, and it was from Reneri that Descartes obtained Scheiner's original observation. What started as a single account was copied, disseminated, annotated, and replicated again within a slowly expanding network of astronomers and natural philosophers, some of whom eventually published on the event.

Mathematicians worked similarly when they sent challenges, problems, and proofs to each other in manuscript form.[111] Often these were competitive efforts to see who could find a solution to a specific problem or a general rule related to number theory. Pierre de Fermat, the French mathematician, was far more comfortable sending challenge problems to colleagues in England and on the Continent, engaging in a closed-circuit correspondence, than he was with printing and publishing his works. A biographer described Fermat's "steadfast refusal to edit and publish his results or to allow them to appear in print under his name" as the singularly most curious thing about him. His mathematical peers would have agreed.[112] In 1657 he sent mathematicians two problems in particular that vexed them for some time, something he enjoyed doing in an almost playful spirit. Even when Englishmen like John Wallis and William Brouncker came up with solutions, they maintained that Fermat's problems lacked any real utility. Wallis told fellow Englishman Kenelm Digby, "I do not think the matter to be of such consequence (for to what end?) that it deserves research in detail."[113] Then again, they could seldom help themselves from giving the problem a try. Descartes replied to Mersenne similarly when sent a math question about numbers and their divisors: "To this, I have nothing to say, because I do not know it and never had any desire of knowing it."[114] And yet, when the problem came up later in correspondence and Descartes realized that several mathematicians, including Fermat, had offered new solutions, he quickly engaged with the challenge, submitting his own ideas in a letter.

Critiques of published works also circulated in manuscript form, since it was impractical to print responses to every scientific treatise that appeared in the booksellers' stalls. Thomas Merry, an algebraist from Leicestershire, critiqued a mathematical treatise of Jan Hudde in a manuscript entitled "Invention and Demonstration of Hudden's Rules for Reducing Equations." John Collins noted in a letter that Merry "hath lent me the Manuscript which is somewhat large, and is willing that any man that pleaseth should have a Coppy thereof or Notes thereout."[115] At least one scribally produced copy was requested by John Wallis, while Collins retained the original.[116] All of Merry's work, in fact, was produced in manuscript form. Upon his death his widow offered them to the Royal Society, in the event they wanted to consider having them printed.[117] We might dismiss Merry's example given that he was a relatively minor figure in the mathematical world, but similar examples can be seen in more prominent figures. Consider Gilles Personne de Roberval, a French mathematician. He had, according to Thomas Hobbes, a peculiar and rather infuriating way of operating: "Whenever people publish any remarkable theorem they have discovered, [Roberval] immediately announces, in papers which he distributes, that he discovered it first."[118] The manner in which Roberval asserted his priority is noteworthy. Rather than print a response, he circulated letters, summaries, or even notes taken by his pupils in lectures.[119] It also reminds us how print was influenced by, and influenced, issues of priority.

Manuscript and letter exchanges were critical to establishing priority, what we would today couch in terms of intellectual ownership. The public nature of epistolary networks could leave authors vulnerable to intellectual theft, even if that threat was mitigated by the slower scribal process. Casting around for ways to protect their work, some authors sought protection within the manuscript exchange by employing ciphers or anagrams, both of which countered the openness of publication. Ciphers typically involved an algorithm that substituted one symbol for another across a text. Following the rule, whatever that happened to be, would unlock the meaning of the otherwise unintelligible jumble. Conversely, anagrams entailed the random rearrangement of letters in a word or phrase that, given the number of characters involved, precluded decryption. Anagrams therefore allowed a solution to a problem to be published but not revealed, and they were used with increasing frequency in the early modern period. It was effectively impossible for a reader to determine the solution to an anagram unless the author elected to explain it, but the existence of a solution was undeniable. Anagrams were deployed by scientists for various reasons, but most frequently to make public a theory that was not fully fledged, what Mario Biagioli called "a pre-dated birth certificate for a claim."[120] The author might require more time for gathering supporting evidence, preparing a publishable manuscript, securing privileges, or obtaining licenses. Whatever the reason, the anagram assured priority through publication without risking theft or premature criticism. Several examples illustrate the ways anagrams were used, either in a public forum that preceded print or in a published work. Galileo used an anagram to announce his discovery of the phases of Venus in 1610. He sent a letter to Giuliano de' Medici, Tuscan ambassador to Prague, with the intention of having it forwarded to fellow astronomer Johannes Kepler. The anagram's solution was a Latin sentence: *Haec immatura a me iam frustra leguntur oy*, meaning "These unripe things are now read by me in vain, oy."[121] When the letters of the Latin phrase were rearranged, they revealed the alternative Latin sentence *Cynthiae figuras aemulatur mater amorum*, or "The mother of love imitates the forms of Cynthia." By this Galileo meant that Venus, the mother of love, has phases that act just like those of our moon, Cynthia being a reference to Diana, goddess of the moon. Galileo's idea was therefore concealed while simultaneously being shared.

Christiaan Huygens took the anagram one step further by actually publishing it. Huygens began his career as an astronomer in the 1650s by devoting himself to both the technical and observational aspects of the discipline. He built his own telescope and began compiling a series of observations of Saturn, a heavenly body whose variable appearance perplexed astronomers. On March 25, 1655, he started following a small star in the vicinity of Saturn, and within three months Huygens had established the period of revolution for this loyal star at sixteen days, four hours. As Galileo had done before, Huygens sent letters to friends, including John Wallis at Oxford and Gottfried Aloys Kinner in Prague. These letters contained the anagram *Admovere oculis distantia sidera nostris, vvvvvvvccccrrhnbqx.*[122] The initial phrase comes from Ovid, "They brought the distant stars closer to our eyes," but the solution to the anagram involves combining the quotation's letters with the random letters at the end. This yields Huygens' solution: *Saturno luna sua circunducitur diebus sexdecim horis quatuor,* or "A moon revolves around Saturn in 16 days and 4 hours." This anagram—a private, epistolary declaration of a concealed fact—allowed Huygens to claim priority for an idea among a select group of astronomers without moving his ideas into the public domain. It also afforded him more time to confirm his theory, which he did in January 1656 when Saturn reappeared. At that point, Huygens used the twenty-three-foot telescope he had constructed with his brother to confirm his theory. The sharper images of this instrument allowed him to speculate further as to the planet's ambiguous shape.[123] For his part, Wallis appreciated being sent both the anagram and its solution, and responded by revealing the solution to his own anagram about Saturn, along with sketches of the planet's appearance illustrating his observations.[124]

Huygens printed his observations of Saturn's moon in the small booklet *De Saturni Luna.* In addition to describing the moon (Titan) and establishing its orbit, Huygens promises the reader that his complete "system of Saturn," by which he means his theory that Saturn has rings, will be made public after he completes additional observations. In the meantime, he writes, "it seems useful to consign the essentials [of the system] to the following anagram, such that, if perhaps anyone believes to have found the same thing, he has the time to make it known and he will not be said to have taken it from us, nor we from him."[125] The anagram reads, aaaaaaacccccdeeeeehiiiiiiiillllmmnnnnnnnnnnnooooppqrrsttttuuuuu.[126] Once solved, it declares, "Saturn is encircled by a thin, flat ring, nowhere touching, inclined to the ecliptic." Unlike the anagram Huygens sent Wallis and others in an earlier letter, this one was semipublic. It appeared in a printed pamphlet circulated only by Huygens and not available from booksellers. As Huygens describes it, the treatise had been "printed" (*imprimée*), but not "published" (*publié*). It therefore told a selective readership that a secret existed but did not reveal the underlying theory. It was not until his more complete *Systema Saturnium* was published in 1659 that Huygens revealed the above anagram and detailed his full theory of Saturn's rings.

The cipher allowed him to print and publish, but not to make public, his ultimate theory.

It is impossible to talk about scientific authors' attitudes toward print without considering the issue of profit, and here, too, anagrams had a role to play. It was obviously important for mathematicians and natural philosophers to receive credit for their discoveries. Scholars have examined the priority disputes that embroiled many in the early modern scientific community.[127] These disputes could be more than academic, as discoveries and inventions, particularly of instruments, were increasingly parlayed into financial gain.[128] Thus, when Huygens used an anagram in 1675 to describe his newly developed spring watch, he was protecting more than priority but a potential patent as well. Oldenburg, who received Huygens' letter containing the cipher, read it aloud at a Royal Society meeting but could not reveal the actual mechanism. One month later Huygens sent the solution to the anagram, explaining how his watch worked, and Oldenburg published the entire explanation in the *Philosophical Transactions* later that year. Huygens worried about obtaining patents for several of his timekeeping inventions, but his concerns paled in comparison to those of Robert Hooke, whose reticence about publishing is well established. He worried that dissemination of his ideas and inventions would allow others to profit from them before he did. In 1676 he published one of his Cutlerian lectures, a work on helioscopes, and appended to it were several anagrams. The first deals with Hooke's efforts to engineer a new kind of dome for St Paul's Cathedral. As Lisa Jardine has shown, Christopher Wren's design for the dome relied on the applied mathematics he developed in collaboration with Hooke. The anagram Hooke included in *A Description of Helioscopes* contained the critical insight of the mathematical catenary that, "as it hangs in a continuous flexible form, so it will stand contiguously rigid when inverted."[129] The second anagram, *ceiiinosssttuu*, contains the key to Hooke's famous spring law (*ut tensio sic vis*, or "as the extension, so the force"). Hooke revealed the anagram two years later in another publication, *De Potentia Restitutiva, or Of Spring*.[130] The last anagram promised "a new invention in Mechanicks of prodigious use, exceeding the chimeras of perpetual motions for several uses," and a reassembly of its letters states, "The vacuum left by fire lifts a weight."[131] Hooke's anagrams, which frequently appeared in published works, allowed him to take advantage of print while still retaining control over his ideas.

One final example of the role anagrams played in print culture comes from Isaac Newton, who in 1676 wrote to Henry Oldenburg about his method of fluxions. This letter is known as the "Epistola Posterior" and is the second of two missives that Oldenburg was to forward to mathematician Gottfried Liebniz in Hanover. Newton, however, was not prepared to fully reveal his method, telling Oldenburg, "The foundation of these operations is evident enough, in fact; but because I cannot proceed with the explanation of it now, I have preferred to conceal it thus: *6accdæ13eff7i3l9n4o4qrr4s8t12ux*."[132] Newton's anagram would resolve into an explanation of the fundamental theorem of calculus.[133] Later in the epistle he uses a second anagram, further explaining his method of fluxions, albeit within a cipher. This was sent to Oldenburg, who eventually had it delivered to Leibniz. But Newton quickly became concerned about whether, and when, his ideas would become public. Writing to Oldenburg just two days after he sent the "Epistola Posterior," he asks that Oldenburg fix some errors in the original letter, and he urges the secretary, "Pray let none of my mathematical papers be printed without my special license."[134] In an era when letters were inherently public documents—especially those sent to Oldenburg—Newton was insistent that his letter, and its contents, be kept out of the broader correspondence networks. Just two weeks later he wrote to John Collins:

> You seem to desire yt I publish my method & I look upon your advice as an act of singular friendship, being I believe censured by divers for my scattered letters in ye *Transactions* about such things as no body els would have let come out without a substantial discours. I could wish I could retract what has been done, but by that, I have learnt what's to my convenience, wch is to let what I write ly by till I am out of ye way.[135]

Years later John Wallis would plead with Newton:

> I wish you would print the two large Letters of June and August [*sic*] 1676. I had intimation from Holland, as desired there by your friends, that something of that kind were done; because your Notions (of Fluxions) pass there with great applause by the name of Leibniz's Calculus Differentialis . . . You are not so kind to your reputation (& that of the Nation) as you might be, when you let things of worth ly by you so long, till others carry away the Reputation that is due to you. I have endeavoured to do you justice in that point; and am now sorry that I did not print those two letters *verbatim*.[136]

Despite the urging from Wallis, Newton was not moved, and these letters succinctly capture his attitude about publication. He was willing to make ideas known to colleagues, even if in a veiled way, but he remained reluctant to send them to the press. The circulation of his ideas in epistolary form was, for Newton, sufficient for sharing ideas and cultivating productive conversations. And far from being inefficient, it was functional. Leibniz, for example, took a trip through England, lodging with Collins in London. Leibniz took advantage of the visit to copy some of Collins' mathematical papers, the originals being "scarce legible," he said.[137] These he later shared with Newton. Likewise, Collins took some pages of Newton's work, along with some from Leibniz, and sent them to another mathematician, Thomas Baker. Collins tells Baker, "I hope speedily to write to you, and *gradatim* transmit the whole, which I doubt not but will remove the doubts you meet with in the new doctrine of infinite series, &c."[138] This kind of tightly knit community, which worked outside of print culture, not only speaks to the persistence of a vibrant manuscript and epistolary tradition in the sciences but reveals a preference on the part of these men for publication avenues outside of print.

Early modern scientists considered their own works in much the same spirit as did Horace—"When you are once let out, there will be no coming back"— knowing that to move a work into print meant confronting myriad challenges. It meant concern about censorship, fear of unprepared readers, reluctance to add another treatise to an exponentially growing list of books in publication, and simply an unwillingness to compromise their peace and privacy. These sentiments are expressed throughout the correspondence of early modern scientists, and they remind us of just how fraught the act of printing could be. They also help explain why scientific authors engaged in vigorous exchanges of correspondence and manuscript. Print did not curb the desire to share ideas and findings privately; it enhanced it. In 1676 Isaac Newton found himself engaged in a debate with the Jesuit priest Francis Line about light, color, and Newton's *experiment crucis*. He was growing weary of the argument, however, and wrote to Oldenburg that "if I get free of Mr. Linus's business I will resolutely bid adieu to it eternally, excepting what I do for my private satisfaction."[139] The relationship between scientific authors and the printing press was complex and highly variable. With some understanding of how print was viewed by these authors, we can turn our attention to the ways they attempted to bring print to heel, making it work for them in the best way possible.

Chapter Two

"To the Unprejudiced Reader"

The Rhetoric of Prefaces in Early Modern Science

> I, and all other booksellers, stand condemned by the bad works of others, and for offering reductions on books written in Spanish or translated from the Latin. For thus armed, fools and ignoramuses nowadays know what in former times was extolled by wise men. Even servants can latinize, and you will come across verses by Horace put into the vernacular in any stable.
> —FRANCISCO DE QUEVEDO, *EL SUEÑO DEL INFERNO*

> I have in obedience to your lordship, and the irresistible suffrages of that society over which you preside, resigned these papers to be disposed of, as you think fit: I hear your Lordship's sentence is, they should be made public.
> —JOHN EVELYN, PREFACE TO *A PHILOSOPHICAL DISCOURSE OF EARTH*

Do Not Read This Book: Prefaces to the Reader in Early Modern Science

In 1671, Sir Isaac Newton wrote to Henry Oldenburg, then secretary of the Royal Society of London, with his summary of an optical experiment. It would be published in the society's journal, *Philosophical Transactions*, which enjoyed limited circulation among its members. In the letter, Newton expresses gratitude that he could share his ideas in the journal "instead of exposing discourses to a prejudiced and censorious multitude, (by which means many truths have been baffled and lost) I may with freedom apply myself to so judicious and impartial an assembly."[1]

At roughly the same time one finds Newton's contemporary, the Dutchman Christiaan Huygens, expressing similar thoughts. Huygens published works in astronomy, natural philosophy, mechanics, and mathematics. He exemplified the new intellectual paradigm of the printing era, when thinkers throughout Europe exchanged books, pamphlets, and printed ephemera across great distances.[2] Yet Huygens was unequivocal in his disdain for the printing process, for many of the reasons outlined in the previous chapter. He lamented time and again the errors introduced to his texts by printers, and he struggled with the notion that his works would be sold to a public unwilling or unprepared to accept new ideas. In the preface to his final work, *Cosmotheros* (1695)—published posthumously—he writes:

> I could wish indeed that all the World might not be my Judges, but that I might choose my Readers, Men like you, not ignorant in . . . true Philosophy; for with such I might promise my self a favorable hearing, and not need to make an Apology for daring to vent any thing new to the World. But because I am aware what other hands it's likely to fall into, and what a dreadful Sentence I may expect from those whose Ignorance or Zeal is too great, it may be worth the while to guard myself beforehand against the Assaults of those sort of People.[3]

And with that he launched, reluctantly, into a lengthy justification for his theories, which included the provocative suggestion that there might be life elsewhere in the universe. By the time he penned *Cosmotheros*, Huygens was an experienced author, having published a dozen works in the Low Countries and France. His reticence about publishing was the product of decades of interactions not only with the world of printers and publishers but with readers. As we will see in the next chapter, Huygens did indeed find creative ways to choose his readers, but he was still plagued with a sense of exasperation over the public nature of his works and the response they were likely to elicit. His specific laments are noteworthy: some readers were simply ignorant, their responses based on an inability to understand the technical nature of problem or idea; others suffered from zealousness—ideological or theoretical—that hindered an objective approach to Huygens' work. His preface is not an attempt to avoid such readers but rather an honest acknowledgment that the audience for his work is not the one he would choose.

Scores of scientific texts from the early modern period contain prefatory epistles similar to Huygens', prefaces that seek to discourage certain readers and solicit others. But they also serve to alert readers to the kind of person who would be able to penetrate the contents of the work. Some prefaces are polite, even flattering: "To the Friendly Reader," or "To the Unprejudiced Reader," while others aim at a more general audience—"The Preface to the Vulgar Reader," *vulgar* used here in the contemporary sense of unlearned, or unversed in Latin.[4] Collectively, these prefaces tell us a great deal about authorial attitudes toward print. As the first two chapters have shown, scientific authors did not fully embrace the technology that could rapidly replicate their works. Mindful of its utility, they were equally wary of the way the press could disrupt traditional communication networks.

This chapter investigates authorial attitudes toward print as expressed in prefaces to the reader, those introductory letters that are found in a majority of early modern books. I begin with an examination of prefaces that aim to dissuade certain readers from engaging with a work. Though there is a degree of rhetoric to be unpacked, these prefatory barriers can tell us about the relationship between scientific authors, their intended readers, and the way printing technology tethers one to the other. In the sciences, some authors were quite reluctant about, if not resistant to, publishing their work. While the press facilitated a broader sharing of ideas and fostered increased communication between scientists in England and on the Continent, prefatory letters are constant reminders of the tension between the need to print and the desire to be read. They address the realities of patronage and credit and often celebrate their important work reaching colleagues who could benefit from it. The price for utilizing the press, however, was exposure to a broader readership, which many scientists disdained. Circulation diluted their authority. The democratizing effects of print often, in their view, created as many problems as they solved by widening readership beyond those who could truly master a work's content.

In contrast to prefaces that attempted to weed out certain readers, some reflected a tremendous enthusiasm for print and welcomed a larger, less academic audience. Here is the preface to a 1670 medical polemic, *The Accomplisht Physician*: "To those whom nature hath raised out of a refined mould, and are by their education sublimed to a higher sphere, as the gentry and literate persons of England, *this discourse is in no wise directed*, unless accidentally by a superficial view, they should give themselves the divertisement of admiring the folly, indiscretion, and fond passion of the vulgar, whom moving erratically in a lower region, is the proper task of these sheets, to reduce to a more certain and less planetary motion."[5]

Curiously, the title page to this work identifies its author as Christopher Merrett, Fellow of the Royal College of Physicians and member of the Royal Society. Scholars believe, however, that *The Accomplisht Physician* was actually penned by Gideon Harvey (unrelated to William Harvey), a London doctor who was not a member College of Physicians and whose relationship with them was often antagonistic.[6] Although he was known for his attacks on un-

licensed apothecaries, a position the college itself shared, he also characterized dissection as a form of cannibalism that contributed nothing to medical
therapies. He levied criticism against experimental medicine as practiced
by many licensed physicians, arguing that doctors should not test on their
patients.[7] Harvey's preface to *The Accomplisht Physician* pitches his work to a
readership outside both the medical community and the educated gentry. He
claims, instead, to offer help for the layman who struggled to delineate between charlatan apothecaries, the surgeons, and the legitimate practitioners.
Yet despite that target audience, the preface demeans those very readers—
ascribing to them a folly and unpredictability not seen among the educated.
What Harvey wants, then, is to alert established, licensed physicians that his
work is not for them. It is for the unlicensed physicians, those practitioners
who prey on the vulgar and complicate medical practice. Prefaces such as Merrett's rhetorically pose as "inclusive" by inviting in a wide swath of readers
who can supposedly benefit from a work. Such prefaces raise questions about
what authors hoped to gain from print and how they used print to leverage
new degrees of authority on a topic. In what follows I examine the ways authors from different scientific fields, different social classes, and different back-
grounds positioned themselves in prefatory epistles, using those "addresses to
the reader" to explore the broad spectrum of attitudes toward print technology
among authors of early modern scientific works.

Prefaces: History and Theory

Scholars have long been attuned to the importance of prefatory material in
early modern books, but the foundational text remains Gérard Genette's 1987
Paratexts. In this work, Genette suggests that paratext "is what enables a text
to become a book and to be offered as such to its readers . . . a heterogeneous
group of practices and discourses."[8] He emphasizes the liminal or "fringe"
nature of paratext, "always the conveyor of a commentary that is authorial or
more or less legitimated by the author . . . a zone not only of transition but of
transaction."[9] It is this transactional nature that I am especially interested in,
the site where authorial intention and a reader's expectations are established.[10]
Paratextual elements include the title page, dedications, images, tables, colophons, and indices—loci not always found in the earliest printed books but
that evolved through the incunable period in response to the needs of authors,
editors, publishers, and readers. Genette made a compelling case for considering the power of paratext, and his work has inspired excellent scholarship

on each of these elements, from the interplay between images and text, to the metaphorical power of a frontispiece, to the role of dedications in early modern patronage dynamics. Most germane for this chapter is the paratextual element of the preface that directly addresses the reader. Genette suggests that the preface acts as a threshold (*seuil*) or vestibule through which a reader must pass before arriving at the text itself, a notion that reflects the thoughts of early modern authors such as in John Evelyn's 1664 *Sylva: or a Discourse of Forest Trees*, where he refers to his preface as "a porch of this wooden edifice." And as William Sherman has shown, the porch metaphor was more than a casual architectural reference; it reflected deeper connections between cognitive activity and physical space in the early modern period.[11] Title pages were framed by elaborate columns, and readers opened a book to find text supported by trusses and scaffolding. Likewise, authorial prefaces created a space to mediate between the world of print and the world of readers, a space that encouraged some to advance and others to feel unwelcome.

The practice of addressing a book's readers through prefatory letters emerged soon after the advent of printing, by the end of the fifteenth century. While the earliest printed books relegated any publishing information to the end, where a printer or publisher might be acknowledged in a colophon, commercial pressures soon inspired an organizational shift, so that publishers began to position their names and emblems at the front of a book. Only then did the author begin to "appear" in the work, first as a name on the title page and eventually as a figure who directly addresses the reader. In William Caxton's 1473 *Recuyell of the Histories of Troy*, the first book printed in England, Caxton speaks directly to his readers in the epilogues of each section. At the end of book 3, for example, he explains how his work was produced: "I have practiced and learned at my great charge and dispense to ordain this said book in print . . . [It] is not written with pen and ink as other books be, to the end that every man may have them at once."[12] By the fifteenth century, printers, translators, and authors increasingly inserted a prefatory note to the reader, often after the title page and dedicatory epistle and prior to the beginning of the actual text. A survey of English works printed in the years 1550, 1590, and 1620 reveals that roughly 54 percent of the prefaces to the reader were written by printers, 36 percent were written by authors, and 9 percent written by translators.[13] Occasionally, prefaces from each of these constituents can be found in a single work. A 1585 English translation of Johann Wecker's surgical treatise, entitled *A compendious chyrugerie*, included the following prefaces: "The Translator to the Reader," "The Booke to the Reader," "To the Printer," "The Transla-tor to the Reader" (another one), "To the Translator," and finally, "In Praise of the Book."[14] Such prefatory abundance in a single work was not common, but Wecker's book managed to cover all possible variations on the theme. My in-terest is in the prefatory material of scientific authors, the thresholds through which readers of math, natural philosophy, medicine, and other works had to pass in order to arrive at the latest ideas early modern scientists had to offer.

"*But what do prefaces actually do?*" The question, posed by Jacques Derrida in his seminal work on the dissemination of meaning, is a reasonable place to begin. His query continues, "Oughtn't we some day to reconstitute [prefaces'] history and their typology? Do they form a genre?"[15] Derrida's central concern was the way meanings could be multiplied by readers, something early mod-ern scholars certainly considered. As Walter Ong writes, "Often in Renais-sance printed editions a galaxy of prefaces . . . establishes a whole cosmos of discourse which, among other things, signals the reader what roles he is to assume."[16] Prefaces do, in fact, constitute a genre, one worth considering more carefully if we want to understand the relationship between early modern scientific authors, their intended audiences, and the way their prefaces acted as buffers between scientific ideas and the world of print. The aim of prefa-tory epistles varied widely, as did their tone. On the face of it, as Genette ex-plains, a preface exists to ensure a work is read properly by alerting readers to the text's importance and to the ideal way it should be scanned.[17] Closer examination of prefaces, however, reveals more nuance and variability in au-thorial motivations. Their justifications for publishing might include a desire to avoid piracy of their ideas or hope that their work would benefit others in the field and therefore advance science. They also used the preface to con-textualize their work and position its arguments within a broader debate, as Robert Boyle did in the preface to his treatise on pneumatics: "I have . . . been moved by the envy of some who, receiving my words blindly and with no understanding, have tried to ridicule me in public. So I have decided to pub-lish my findings so all may form an opinion of me and of the work itself."[18] A great many prefaces served as a locus for false modesty, with the author apologizing—assiduously, and often disingenuously—for publishing at all. It had long been considered ungentlemanly to seek publicity for one's work, a circumstance authors attempted to mitigate by offering excuses for appearing in print. Boyle was a master of the trope. He went to great lengths to assure his readers that "intelligent persons in matters of this kind persuaded me, that the publication of what I had observed . . . would not be useless to the world."[19] Indeed, being useful to the world was a commonly claimed motivation for publishing. The preface to a 1669 compilation of recipes by Sir Kenelm Digby, a natural philosopher and alchemist, is a paradigmatic illustration of prefatory verbosity commonly seen in the period: "This collection full of pleasing variety, and of such usefulness in the Generality of it, to the Publique, coming to my hands, I should, had I forborn the Publication thereof, have trespassed in a very considerable concern upon my Countrey-men, The like having not in every particular appeared in Print in the English tongue. There needs no Rhetoricating Floscules to set it off."[20]

Digby's preface was actually penned by his editor, but reference to "rhetoricating floscules" is precisely the kind of verbiage that invited contemporary lampooning. Jonathan Swift's sharp-edged satire of such prefaces is seen in *Gulliver's Travels* when Gulliver writes to his publisher, Richard Sympson, with a lengthy justification for the work. The letter, which assumes the position of a preface to the reader, draws readily on the tropes of the period: "I hope you will be ready to own publicly, whenever you shall be called to it, that by your great and frequent urgency you prevailed on me to publish a very loose and uncorrect account of my travels."[21] Gulliver goes on to criticize Sympson and his cadre of university student assistants for altering the original text in ways that Gulliver finds problematic, a lament that echoes many authorial complaints about the editorial interventions of publishers. "Several passages of my discourse . . . you have either omitted . . . or minced and changed them in such a manner, that I do hardly know mine own work." But Swift's satirization of early modern prefaces hits its stride with Gulliver's litany of complaints and his assertion that he never should have published at all:

> I do in the next place complain of my own great want of judgment, in being prevailed upon by the entreaties and false reasonings of you and some others, very much against mine own opinion, to suffer my travels to be published . . . I find likewise, that your printer hath been so careless as to confound the times, and mistake the dates of my several voyages and returns . . . And I hear the original manuscript is all destroyed, since the publication of my book. Neither have I any copy left; however, I have sent you some corrections, which you may insert, if ever there should be a second edition: And yet I cannot stand to them,

but shall leave that matter to my judicious and candid readers, to adjust it as they please.[22]

Swift's knack for tapping into, and mocking, authorial prefaces is more than humorous; it illustrates how common these prefatory letters and their tropes had become by the early eighteenth century. If we survey prefaces to scientific works in the sixteenth and seventeenth centuries, we can begin to develop a kind of prefatory taxonomy: the apologizing or justifying preface; the contextual preface that aims to place a work within an intellectual tradition; the architectural preface that outlines how the work is built; the prescriptive preface, which instructs readers on how to approach the contents of the work; and finally the exclusionary preface, which articulates who should and should not read it. Underscoring many of these is the author's attempt to mitigate the so-called stigma of print that attached to publishing in this period, particularly in regard to men of high social standing.

The idea of a stigma attached to publishing was articulated originally by J. W. Saunders in his 1951 article on Tudor poetry. Put simply, Saunders argues that gentlemen shunned print.[23] To negotiate this stigma, poets—and by extrapolation other gentlemen authors of the Renaissance—used a variety of tools to cloak their identity or to otherwise displace responsibility for publication, all the while asserting their own modesty and humility. Economic interest and a desire to augment one's reputation were, according to Saunders, out of line with aristocratic values at the time. Rather than publish widely, gentlemen preferred to write for a more intimate manuscript audience, which Saunders suggests was distinct from the audience for a printed book. While Saunders makes some keen observations about aristocratic attitudes toward print, his acceptance, a priori, of a distinction between manuscript and print culture was not born out by subsequent research. Renaissance scholars began to view the dichotomy between manuscript and print as less rigid and more complex. Elizabeth Eisenstein, nearly thirty years after Saunders, saw the transition to print culture as much slower, something that took place gradually over two hundred years after its invention. In such a framework, it makes little sense to talk about authorial attitudes toward print in sharp terms, as if the technology of the press had instantaneously transformed the publishing landscape. Rather, historians from Eisenstein to Adrian Johns have presented evidence that authors from across disciplines had complicated and dynamic attitudes toward both manuscript publication and print. Saunders' stigma, however, did not entirely disappear. Authors in the sciences continued to express hesitation about printing their works.

A more recent examination of attitudes toward printing, by Wendy Wall, uses the lens of gender to illuminate other sources of reluctance on the part of authors, linking masculinity to manuscript in interesting ways.[24] Wall considers what it meant in the seventeenth century to "undergo a pressing," a contemporary term that implied feminization of the author through the assumption of the female, or submissive, sexual position. Specifically, she argues that when a Renaissance man published his work, when the literal impression was made from inked type, he lost his authorial virginity. The metaphor is hardly a stretch, as Wall's extensive evidence demonstrates. In the earliest years of print, the Dominican friar Filippo de Strata put it simply: "Est virgo hec penna: meretrix est stampificata" (The pen is a virgin, the printing press a whore).[25] Well into the seventeenth century, the act of printing was fraught with meaning as the author—emasculated—issued the products of his labor from the press, often with the help of a midwife, or editor. The offspring was a book, a body of writing that Philip Sidney called, somewhat dramatically, "thwarted infanticide."[26] René Descartes, in a letter to Constantijn Huygens, explained that his work *Le Monde*, "would be out already were it not that first of all I want to teach it to speak Latin. I shall call it the *Summa Philosophiae*, to help it gain a better reception among the Schoolmen, who are now persecuting it and trying to smother it at birth."[27] To carry forward Sidney's metaphor, authors not only worried that their scientific efforts would be printed incorrectly, yielding a deformed offspring; they feared for how those works would be accepted by readers. If printing prevented the "infanticide" of an author's ideas, then an uncontrolled audience would leave those ideas improperly nurtured. Prefaces attempted to circumvent some of this misapprehension and ensure that the offspring of an author was properly attended to, rather than neglected. The sixteenth-century writer Thomas Dekker captures the sentiment best: "Readers . . . are not Lectores, but Lictores, they whip books (as Dionysus did boys) where as to Understanders, our libri, which we bring forth, are our Liberi (the children of our brain) and at such hands are as gently intreated, as at their parents: at the others, not."[28]

It was more than concern for their "offspring" that prompted authors to engage readers in a preface; it was also their desire to forestall the crumbling of social distinctions that widespread circulation implied. When a scientific author moved beyond a closed network of communication among peers, they opened themselves up to the possibility that the intrinsic value of their work would decrease. Thus, a preface that identified the "proper" readers for a book could restore a measure of the work's intellectual value and the authority of the scientist who composed it. It is this maneuvering for authority that gave the preface its greatest potential.

Kevin Dunn's study of Renaissance prefaces considers the issue of authority carefully and takes a different view of the modesty topos found in so many of them. He examines the deep ambivalence expressed in many prefaces, which was "especially acute in the seventeenth century scientific writers, between older notions of intellectual activity as the retired musings of the aristocrat and a powerful new definition of that activity as labor, public work for the public good."[29] Dunn, who looks at figures like René Descartes and Francis Bacon, suggests that early modern scientists struggled to repair the seam between private intellectual work and public presentation of that work. Though Bacon managed to rhetorically navigate the two worlds, other scientists remained vexed by the mantle of public authority that printing placed on them. Their desire to "commute between sets of values" is evident in the posturing of their prefaces.[30] Building on Dunn's framework, scholars have explored authorial attitudes toward print in the Renaissance. Martin Elsky, in particular, suggests that our understanding of authorial values is improved if we consider categories of writers in this period: the aristocratic and the professional.[31] The former, consisting largely of gentlemen amateurs, wrote for political or social gain. Their works traditionally circulated in manuscript form and targeted specific patrons and small networks of readers. Broadcasting their works was disadvantageous, and thus authorship through print was perceived as antithetical to the ideals of courtier.[32] Alternatively, certain men of letters began to identify themselves as professional authors whose vocation depended on the printing press. Wide circulation was, for them, a means to professional ends. Critically, Elsky traces the evolution of attitudes toward publication to show how authority became increasingly tethered to printing, rather than diminished by it. To put one's ideas into print form, as opposed to leaving them as manuscripts, imbued them with permanence that could propel an author and his or her reputation into the future. Francis Bacon, an exemplar of Elsky's professional class who embraced print, asserted as much in his *Advancement of Learning*, claiming that "the images of men's wits and knowledges remain in books, exempted from the wrong of time, and capable of perpetual renovation."[33]

If we apply Elsky's framework to early modern scientists, a hybrid form of authorial figure emerges. Authors of early modern mathematical, astronomical, and natural philosophical texts did not consider themselves, by occupation, professional writers. They viewed themselves as gentlemen for whom intellectual pursuit was paramount. But as such, they often sought patronage in some form, particularly in the years before the emergence of scientific societies. Galileo is an excellent example. His deft use of print to cultivate patrons allowed him to advance his career from a middling university professor to an esteemed court mathematician. His printed publications were integral to his engagement with the economy of scientific support in the period.[34] But Galileo understood his role and identity vis-à-vis the press. He sought to explain natural phenomena, not to please an audience or earn a living through his writing. Publication was simply a necessary step to communicate ideas and to secure priority for discoveries.[35] In the pre-print era, when manuscripts were exchanged among scholars in a field, this was in some ways simpler.[36] Circles of correspondents had long existed, bound together by the parchment shared among members. Printing pulled at the bonds of those closed networks and ruptured a centuries-old system of communication. We saw in the previous chapter that although these epistolary communities did not disappear—correspondence around Oldenburg, Mersenne, and Leibniz flourished in the seventeenth century—most authors accepted the need to move beyond them and print their works. The prefaces they included in these works reflect some of the tensions around that shift, and they provide us with a window into the attitudes and aims of scientists who authored them.

The Early Modern Scientific Preface: Case Studies

A good place to begin this survey is with those that might be termed "exclusionary prefaces." These prefaces acted as vestibules built to discourage readers, rather than invite them in. In such anterooms, the chairs were uncomfortable and the space inhospitable. As Giovanni Borelli writes in the preface to his treatise an animal anatomy, "All men are called to examine and read this divine book and none is excluded from its vision. However, it is not given to everyone to enter its sanctuary. It is not allowed to everyone to read and understand the secret sentences which are written in the living characters of this book."[37] We can also consider the examples of Johannes Kepler, whose astronomical work transformed the field. He posited the theory of elliptical orbits for the planets and—more importantly—a physical approach to astronomy that rejected the ancient, mathematical tradition. That is to say, Kepler believed his understanding of the heavens, presented mathematically, described what was actually up there, whereas traditional mathematical models did not concern themselves with correlating to reality. As scholars of Kepler's work have shown, this view of the universe, which brought together astronomy and physics, was "the most important conceptual change in science during the period."[38] It was an idea that challenged long-held beliefs about planetary motion and established a paradigm for conceptualizing the universe that would prove foundational for Isaac Newton's later work. Kepler understood, as astronomers before him had learned, that such innovation was more likely to cause conflict and censure than to cultivate debate, at least among laypeople. His prefaces to the reader, therefore, had to justify his approach to astronomy and to delineate his audience.

In 1602 Kepler wrote a treatise entitled *On Giving Astrology Sounder Foundations* (*De Fundaments Astrologiae Certioribus*). His preface to the reader begins with a defense of this area of science, which demands a level of rigor: "These are worthy studies and fortitude too is required in prosecuting them, fortitude which strengthens the mind against the foolish opinions of the crowd and despises perverse judgments."[39] Kepler acknowledges that putting a work into print exposes those ideas to a range of readers for whom it is not suited. "If anyone objects that this work is addressed to the public at large, who are least likely to draw profit from it, I should like to ask him to bear in mind that by no other way than publication could we reach out to the educated persons hidden here and there among the crowd" (230). Manuscript communication can be effective, but Kepler acknowledges that it is almost wholly inefficient if one desires an audience beyond a known and established community. He continues, "I hope the right minded will not suspect me of anything underhanded when, with the best intentions, I publicly speak to the crowd . . . on the subject of what is to come. As for unfair judges, who do not go deeply into matters even of the greatest import, and merely mock, if they do not leave untouched this humble professional service, so open to ridicule, I shall follow the poet's advice and turn the unseeing back of my head to them" (231).

The central paradox is addressed here by Kepler: there is no way to contact the educated readers in Europe without resorting to print, and yet it is through print that Kepler finds himself exposed to a public unprepared for what he has written, even among the educated who could read his Latin text. As we saw in the introduction, Kepler begins his *New Astronomy* of 1609 lamenting the unprepared audience his work would likely get. He also discusses the difficulties of writing about mathematics and trying to explain complex ideas clearly, without becoming too verbose. Kepler is keenly aware of the challenges he confronted when putting a work into print, and he distills the problem faced by so many scientists of the period: they needed the printing press to reach the learned but loathed the printing press for the wide net it cast.

To deal with this, Kepler attempted in his preface to prepare readers with different skill sets for his work. Some, he said, would be more skilled in the physical sciences and find his tables to be "more tangled than a Gordian Knot." For them he included a prefatory synopsis. Other readers might be astronomers coming from a different paradigm, such as those who subscribe to the astronomical views of Tycho Brahe, and for them Kepler provides specific arguments in the preface to demonstrate the superiority of a Copernican view. Ultimately, Kepler hoped that the preface to his *New Astronomy* would frame the work for the physicists, astronomers, and geometers, each constituency bringing different skills to the treatise, each requiring a different kind of proof.[40] Kepler went to great lengths to accommodate varying epistemological approaches to astronomy, but he had less patience for readers who challenged his ideas with words of Scripture. He walks them through myriad Old Testament verses to demonstrate why they should not be relied on to describe the structure of the universe. And finally, he arrives at a section that is called, in the margin, "Consilium pro Idiotis" ("Advice for Idiots"). As Richard Oosterhoff has noted, *idiota* at the time referred to those who lacked Latin grammar, and Kepler seems to be working from that definition, albeit with some augmentation.[41] Here his prefatory priming turns into an unequivocal warning: "But whoever is too stupid to understand astronomical science, or too weak to believe Copernicus without affecting his faith, I would advise him that, having dismissed astronomical studies and having damned whatever philosophical opinions he pleases, he mind his own business and betake himself home to scratch in his own dirt patch, abandoning this wandering about the world."[42]

What Kepler and other *illuminati* of the scientific world endeavored to do through their prefaces was set the bar high enough to discourage unqualified readers. A similar example is seen in the work of Isaac Newton. In his 1687 *Principia Mathematica*, Newton applied mathematics to the system of the universe, thereby describing—among many other things—universal gravitation. Two chapters—or what he calls books—build his case. Book 1 offers Newton's mathematical treatment of bodies in motion in the absence of resistance; book 2 treats bodies in motion through a resistant medium. It was the third book, however, "On the System of the World," that brought Newton's ideas together in a brilliant exposition of how gravity acts in the universe. Book 3 is the summation of the entire treatise and among the greatest expositions of physics in the history of science, which makes Newton's preface to that section all the more interesting. He explains not only what he hopes to accomplish in the book, but also how he casts the work at a specific readership:

> Upon this subject I had, indeed, composed the third book in a popular method, that it might be read by many; but afterward, considering that such as had not sufficiently entered into the principles could not easily discern the strength of the consequences, nor lay aside the prejudices to which they had been many years accustomed, therefore, to prevent the disputes which might be raised on such accounts, I chose to reduce the substance of this book into the form of Propositions (in the mathematical way), which should be read by those only who had first made themselves masters of the principles established in the preceding books.[43]

Newton shared with others what Rob Iliffe called "the genteel distaste for print," but what he demonstrates in the *Principia* are the measures a scientific author might take to ensure a fair, if not positive, reception. Few readers had the expertise to understand the work; in Newton's mind, these should be the only people given a chance to respond to it. Thus, the door to the interior of the work was effectively locked by the author keen to avoid an ill-prepared reader unable to make sense of mathematical or philosophical complexities.

Few other authors in natural philosophy, mechanics, or mathematics so expressly closed off their work to a general audience. Their prefatory remarks were less exclusionary and more didactic, providing the reader with an overview of the work and instructions on how best to navigate it. These prefaces ranged from general advice on how to prime one's thinking to engage with a treatise to more detailed techniques one should use while reading because, as Boyle observes, not all things are written for all readers.[44] In 1638 John Wilkins, a member of the Royal Society, wrote *The Discovery of a World in the Moone*, a book that addresses the probability of life existing on the lunar landscape. In his preface to the reader, Wilkins issues a plea for objectivity: "Let me advise thee to come unto [this idea] with an equal mind, not swayed by prejudice, but indifferently resolved to assent unto that truth which upon deliberation shall seem most probably unto thy reason."[45] He then prepares the reader for what is to come, cautioning that he is offering an idea that is only probable, and that the reader seeking certainty will not find it: "You must not look that every consequence should be of undeniable dependence, or that the truth of each argument should be measured by its necessity."[46] This is more than a preface; it is a primer on the scientific method. Wilkins understood that some readers were unaccustomed to hypotheses, unfamiliar with the act of summoning evidence in support of a theory. He closes the preface with a more philosophical view, one that summons the spirit of Francis Bacon while implicitly prescribing the proper audience for his work: "It must needs be a great impediment unto the growth of sciences, for men still so to plod on upon beaten principles, as to be afraid of entertaining any thing that may seem to contradict them. An unwillingness to take such things into examination, is one of those errors of learning in these times."[47] Wilkins begins his arguments in favor of the existence of extraterrestrial life in a carefully erected series of propositions. The first squarely reinforces the approach he recommends in his preface, that the strangeness of an idea is not sufficient reason for its dismissal. Surely, he suggests, ideas about nature have been presented that, on first take, seemed nearly inconceivable but were subsequently proven to be true. Likewise, ridiculous ideas have become enshrined for no reason other than blind consensus. "I shall give an instance of each," Wilkins writes, "so I may the better prepare the reader to consider things without a prejudice."[48] Wilkins' didactic preface, his reminder that the advancement of science depends on people's willingness to entertain ideas that seem to contradict what they know, was echoed by other scientists in the period, in part because it was increasingly necessary.

A similar example is in Sir Kenelm Digby's *Two Treatises: Of Bodies and Man's Soul*. Digby's engagement with the manuscript tradition—exchanging letters, treatises, and essays with colleagues—was significant, but he also understood the importance of putting one's ideas into print.[49] In a prefatory letter to his son, he explains that readers of his work need to consider the treatise as a whole, to "draw the entire thread through their fingers and . . . examine the consequences of the whole body of the doctrine I deliver."[50] If they fail to read it completely and carefully, or if they "ravel it over loosely . . . which is the ordinary course of flashy wits who cannot fathom the whole extent of a large discourse," Digby guarantees that they will be unsatisfied with his conclusions. The same warning is echoed in his address to the reader: "I will . . . end this preface, with entreating my reader to consider, that in a discourse proceeding in such order as I have declared, he must not expect to understand, and be satisfied, with what is said in any middle or latter part unless he first have read and understood what goeth before. Wherefore, if he cannot resolve with himself, to take it along orderly as it lyeth from the beginning, he shall do himself (as well as me) right, not to meddle with this book."[51]

If Kenelm Digby provided his readers with general instructions on how to read his work, René Descartes went into much more depth. Prefaces to his work provide us with tremendous insight as to how Descartes envisioned his readership, those he believed should steer clear of his work, and the way astute readers should approach his dense content. Two of his works will suffice as examples, the 1641 *Meditations on First Philosophy* and the 1647 *Principles of Philosophy*.

The *Meditations* were first published in Latin and provided the foundation for much of Descartes' philosophical work dealing with the soul and the existence of God. In his preface to the reader, Descartes explains that his thinking in these two arenas is both novel and complex, and his path to explaining them is "remote from the normal way." He also references the reception of his previous publication, *Discourse on Method*, acknowledging some valid criticisms but noting that "the judgement of many people is so silly and weak that, once they have accepted a view, they continue to believe it, however false and irrational it may be . . . So I do not wish to reply to such arguments here."[52] Instead, Descartes attempted something new with his *Meditations* in an effort to circumvent the poor judgments of a reader. After composing the treatise, which consisted of six individual essays on aspects of metaphysics, he sent manuscript copies to both conservative and more innovative thinkers in the Low Countries and throughout France, requesting feedback from his chosen readers. Soon after, he received letters from these individuals with their responses, thereby recreating the scholastic approach to learning through disputation.[53] Their philosophical objections to his various meditations were addressed by Descartes in a series of replies, which were then appended to the work. When the entire thing was published, a reader would encounter Descartes's original meditation, a particular scholar's response to it (called an objection), and Descartes's reply to that objection. As he wrote in a letter to Mersenne, he was not worried that the *Meditations* would be offensive to theologians; nevertheless, he writes, "I would have liked . . . the approbation of a number of people so as to prevent the cavils of ignorant contradiction-mongers. The less such people understand it, and the less they expect it to be understood by the general public, the more eloquent they will be unless they are restrained by the authority of a number of learned people."[54]

When the *Meditations* finally went to the press of Michael Soly in Paris, Descartes' preface to the reader closed with a frank assessment of the work's intended audience:

> I do not expect any popular approval, or indeed any wide audience. On the contrary I would not urge anyone to read this book except those who are able and willing to meditate seriously with me, and to withdraw their minds from the senses and from all preconceived opinions. Such readers, as I well know, are few and far between. Those who do not bother to grasp the proper order of my arguments and the connection between them, but merely try to carp at individual sentences, as is the fashion, will not get much benefit from reading this book. They may well find an opportunity to quibble in many places, but it will not be easy for them to produce objections which are telling or worth replying to.[55]

Descartes' efforts to obtain feedback on his work before publication offer an interesting and rare view on the reception of his ideas. Early modern historians have increasingly explored texts from the perspective of readers, an understandably challenging approach because it relies on a certain amount of serendipity. The vast majority of readers did not record their reactions to scientific works in any formalized manner. However, through an excavation of marginalia, personal notes, correspondence, and excerpting (or commonplacing), scholars have seen how specific readers interacted with a work, interpreted its contents, and responded to its ideas. What they have found often challenges traditional ideas of how texts—especially in the sciences—were read. In her study of Galileo's *Two New Sciences*, for example, Reneé Raphael discovered that readers often approached Galileo's work quite differently than historians have assumed; they brought to the text questions and problems scholars did not expect. Thus, Raphael has opened the door to a plurality of readings, some based on what scholars know about Galileo and his theories today, and others predicated on what his contemporary audience read in his work. In one instance, Raphael looks at how an "ideal reader" like the Italian natural philosopher Giovanni Battista Baliani annotated his copy of *Two New Sciences*.[56] She compares that to readings by English natural philosophers Seth Ward and Christopher Wren. The latter employed the Baconian methods of "note-taking via commonplace headings," so that information from Galileo's book could later be collated with new data from their own experiments. Through such annotations, Raphael reverse engineered the responses readers had to specific works. In a similar fashion, Descartes' *Meditations*—with the replies and objections to those replies—affords us the opportunity to consider multiple readings.

We see something similar in the French edition of Descartes' *Principles of Philosophy*, published in 1647. There he includes a preface that he casts as the "Author's letter to the translator of the book which may serve as a preface." That is to say, Descartes wrote a preface for his work, something he did not do for the 1644 Latin edition, but he avoided direct communication by passing the preface off as a letter to a third party. His motivations for this are not immediately clear, though Kevin Dunn suggests that Descartes hoped to preserve some of his private authority while acting as a public spokesman for his philosophical ideas.[57] Descartes viewed books much as he did correspondence, as tools that allowed readers to have a "conversation" with the author, and he used his prefatory remarks to frame that conversation: to explain the topics, to establish the logical ground rules, and to suggest an authority resident in the pages.[58]

The preface to his *Principles* explicitly advises readers "about the way to read this book."[59] But Descartes goes well beyond other scholars' admonitions to approach the work with an open mind. He offers *very specific* advice about reading his treatise:

> I should like the reader first of all to go quickly through the whole book like a novel, without straining his attention too much or stopping at the difficulties which may be encountered. The aim should be merely to ascertain in a general way which matters I have dealt with. After this . . . he may read the book a second time in order to observe how my arguments follow. But if he is not always able to see this fully, or if he does not understand all the arguments, he should not give up at once.[60]

He goes on in some detail, advising the reader to use a pen and mark the places where he struggles and to revisit difficult sections after reading the book through. Studies have shown that approximately 50 percent of the extant books from this period contain marginalia.[61] Readers were doing precisely what Descartes asked, annotating as they read, leaving the traces of reader response that historians welcome. Finally, Descartes assures the reader that three passes through the text should be sufficient to get the material; at most, a fourth reading will suffice, even for the more complex matters. With such prescriptive instructions, Descartes comes across as a supportive guide—almost a tutor—but at the same time he erects a partition between him and the reader, demanding more than a dilettante's attention to the text. There is hardly anyone, he says, who cannot grasp his positions if sufficient attention is paid to their construction. He purports to be familiar with a range of thinkers and mental capabilities but insists that there are "almost none that are so dull and slow as to be incapable of forming sound opinions or indeed of grasping all the most advanced sciences, provided they received proper guidance."[62] Rhetorically this assessment allows him to preemptively filter out not only readers who cannot navigate his ideas but also those who might disagree with his philosophical conclusions. To disagree, he suggests, is to simply misunderstand.

Descartes continues in his preface to explore his overarching methodological approach, highlighting one of the primary perils of print, which is that anybody can take advantage of the technology. For Descartes, print *can be* used to great effect, allowing a scholar to share new philosophical tools; however, he readily acknowledges the inherent problems: "I am well aware that there are some people who are so hasty and use so little circumspection in what they do that even with very solid foundations they cannot construct anything certain. Since such people are normally quicker than anyone else at producing books, they may in a short time wreck everything that I have done."[63]

It was printing that allowed Descartes to share his ideas, but it was printing that could so quickly undo the philosopher's efforts. Setting aside the obvious value Descartes placed on his own work, his acknowledgment that ideas— once published—become legitimized in the minds of readers is notable. "I must also beg my readers never to attribute to me any opinion they do not find explicitly stated in my writings."[64] Descartes is gesturing at what we might call typographical epistemology, the idea that once the type has been set and the pages pressed, the ideas contained in drying ink gained authority. Impressions confer veracity.

Where Descartes was dealing with a philosophical system that he wanted to be considered holistically, other natural philosophers presented theories and ideas that could withstand discrete, or fragmented, study. In the preface to his 1664 work, *Some considerations touching the usefulness of experimental natural philosophy*, Robert Boyle explains to the reader that certain essays might not be for everyone. Readers are advised to consider the work in discrete sections, rather than as a whole, if it facilitated their understanding. "Those readers, that either cannot relish, or at least desire not any thing, but what is merely physiological, may, thus advertised, pass by the former part of this treatise, and content themselves to read over the latter," he writes. Despite this acknowledgement that skimming may suit some readers, he does note that those who take in the entire book "will not perhaps think their labor lost." But Boyle understood his audience was heterogeneous, and his advice to the reader reflected that.[65] Boyle also used his prefaces to the reader to assert priority on a variety of discoveries and to establish a record of his publishing as a bulwark against potential accusations of plagiarism. As Michael Hunter has shown, Boyle frequently adopted an apologetic tone in his prefaces in a manner consistent with tropes of authorial modesty, but such a posture was matched by his use of the preface for polemical purposes, decrying the unattributed use of his work by others.[66]

By contrast, there were authors who felt little need to prepare their readers for the material to follow, nor did they attempt to erect barriers to their work. Instead, they assumed from the start that a subset of their audience would read their ideas and still be antagonistic. These authors might choose to rhetorically gird themselves for the coming criticism or to preemptively isolate critical readers, as if to quarantine their rebuttals. Margaret Cavendish, Duchess of Newcastle and well-known intellectual of the seventeenth century, is a useful example. Cavendish wrote poetry and plays, essays, and works of philosophy and science, and although she praised the existence of print and expressed no false humility about taking advantage of the medium, she had no patience for those who did not like her work or who did not agree with her ideas. In a prefatory epistle to *The Worlds Olio* she notes that "several Censures can never enter to vex me with wounds of discontent . . . [T]hose that do not like my book, which is my house, I pray them to pass by, for I have not any entertainment fit for their palates."[67] In a separate preface Cavendish sets out her expectations of the reader: "I Desire those that read any of this Book, that every Chapter may be read clearly, without long stops and stays." For her, a smooth and continuous reading was the only way to capture the author's intention: "Writings if they be read lamely, or crookedly, and not evenly, smoothly, & throughly, insnarle the Sense." She goes on to identify two kinds of reader: "the one that reads to himself and for his own benefit, the other to benefit another by hearing it; in the first, there is required a good Judgement, and a ready Understanding; in the other, a good Voice, and a graceful Delivery; so that a Writer hath a double desire, the one that he may write well, the other, that he may be read well; And my desire is the more earnest, because I know my Writings are not strong enough to bear an ill Reader."[68] Cavendish's self-deprecating tone, the idea that her work could not stand up to poor reading, resembles the false modesty of her male counterparts, but unlike those men Cavendish places significant emphasis on her gender. She was hardly naïve about the implications of a woman, even one of high social standing, writing on natural philosophy. In the privacy of his diary, Samuel Pepys describes her as "a mad, conceited, ridiculous woman," yet like many others he clamored to see her when she came to London.[69] In several works she articulates a conviction that women were inherently weaker scholars than men, and as an autodidact she expresses insecurity about her lack of formal education. Referring to her 1668 *Grounds of Natural Philosophy* as her newborn child, Cavendish acknowledges her resistance to any outside help: "It is so commonly the error of indulgent parents to spoil their children out of fondness, that I may be forgiven for spoiling this, in never putting it to suck at the breast of some learned nurse, whom I might have got from among [university] students, to have assisted me; but would, obstinately, suckle it myself, and bring it up alone, without the help of any scholar."[70]

Despite the maternal metaphor, she could also be playful, if not vexing, about her gender identity. When she visited the Royal Society in 1667—the first woman to do so—she played the provocateur by dressing outrageously and androgynously, treating the visit as a kind of carnival pageant that, it seems, most of the fellows enjoyed.[71] We should therefore view her prefatory modesty as drawing on the rhetorical practices of the day but unleashing an even greater level of satirical intensity than those of her male peers.

Like other scientific authors, Cavendish shared her frustrations about working with printers and publishers. She decries the many typographical errors that found their way into a printed work, "for by the false printing, they have not only done my Book wrong in that, but in many places the very Sense is altered . . . so that my Book is lamed by an ill Midwife and a Nurse, the Printer and Overseer."[72] Her offspring may have been harmed in the print shop, but an even greater concern was the quality of the readership that would consider the work. In addition to her comments on the kinds of reader she wanted and the problems such a reader might encounter, Cavendish used her prefatory epistles to insulate herself from the criticisms that would invariably be prof-fered. In the opening preface to *The philosophical and physical opinions*, she preempts a range of criticisms she expects will be levied against her:

> But those that make these and the like idle objections against me either have not read all my epistles, and the rest of my books or understands them not, but that is not my fault, but their unjust natures, to censure and condemn before they examine or understand; Nay, they do in some things falsely accuse, and maliciously break out of some of my epistles some parts to throw against me which is most base and cruel to dismember my book tormenting it with spiteful objections misforming the truth with falsehood: but those that have noble and generous souls will believe me, and those that have base and mechanick souls, I care not what they say . . . for I have observed that the ignorant, and mali-cious, do strive to disturb, and obstruct all probable opinions, witty ingenuities, honest industry, virtuous endeavors, harmless fancies, innocent pleasures, and honorable fames although they become infamous thereby.[73]

Cavendish excoriates not just the unlearned or careless reader; she implicates other natural philosophers, such as those of the Royal Society, who neglect to consider her work carefully. She understands the inherent bias against her work, coming from a woman, someone untrained in the schools, someone who sits outside of the accepted scientific community: "I verily believe, that ignorance and present envy will slight my book," she writes, "but understand-ing will remember me in after ages . . . I had rather live in a general remem-brance, than in a particular life."[74]

While Cavendish articulated a defensive posture based on her gender, oth-ers found that prefatory bulwarks were useful when addressing controversial topics, such as those dealing with natural magic. Italian alchemist and natural philosopher Giambattista della Porta published an English translation of his *Natural Magik* in 1658. The treatise, originally issued in 1558, was a popular exploration of everything from medicine to optics, magnetism to geology. With the work, della Porta influenced a generation of European scholars interested in understanding and manipulating nature through work in the secular tra-dition of the magus. Yet he understood that a wide audience was problematic. He quotes Plato: "They seem to make Philosophy ridiculous, who endeavour to prostitute Her Excellence to prophane and illiterate Men."[75] He repeats this sentiment in his preface to "Studious Readers," written thirty-five years after he first published the book: "I was somewhat unwilling to suffer it to appear to the publick view of all men . . . for there are many most excellent things fit for the worthiest nobles, which should ignorant men (that were never bred up in the sacred Principles of Philosophy) come to know, they would grow contemptible, and be undervalued." Indeed, della Porta claims to have omitted nothing from his account, though he says he has "veil'd by the artifice of words" those things that are most excellent and magnificent. Information that might be used to cause harm he has likewise "written obscurely; yet not so, but that an ingenious reader may unfold it, and the wit of one that will thoroughly search may comprehend it." He acknowledges, as others do in their prefaces, that "there will be many ignorant people, void of all serious matters, that will hate and envy these things," but he makes a plea for objective reading. "Remove all blindness and malice, which are wont to dazzle the sight of the mind, and hinder the truth; weigh these things with a right judgment." Della Porta's preface came in response to attacks on him personally and on his field of natural philosophy, wherein the mysteries of nature were explored and nature's powers harnessed. But his desire to lift the veil on his findings was also a response to a broad clarion call from the scientific community, one that urged openness over secrecy.

A more pointed critique of potential readers is seen in the prefatory epistle to Robert Boyle's *A Defence Of the Doctrine touching the Spring and Weight Of the Air* (1662). This was Boyle's response to the challenges posed by the Jesuit priest Francis Line. In the preface to the reader Boyle summons a measure of authority when he references "excellent mathematicians" and "eminent naturalists" who support his ideas. He also assures the reader that a number of virtuosi originally questioned his ideas but says that they were convinced after careful reading of his experiments. The reader is implicitly held up against these learned men, the judgment of the former pressured by the conclusions of the latter. Boyle closes his preface by explaining why he cannot, and will not, respond to each and every argument against his theories. Certainly, he will address queries that he deems legitimate, perhaps even in a subsequent treatise, "yet I would not have it expected that I should exchange a book with every one that is at leisure to write one against a *Vacuum*, or about the *Air*."[76] The reader can expect only serious replies to earn Boyle's attention. Thus, he says, a reader should not take Boyle's silence on an issue as a concession. Moreover, he claims that by ignoring weak explanations or theories that challenge him, he might "give unbiased and judicious readers a caution to allow as little of advantage to the writings of my adversaries upon the account of their being unanswered by me, as if I were no longer in the world."[77] It is a decisive assertion of authority: if I do not respond to an attack, you can be assured that the attack is not worthy of your time.

Similarly, we can look at the 1671 medical treatise of French anatomist Jean Riolan. In the preface to the reader of this anatomical work, Riolan explains how the book should be read and how the reader should integrate the notations and accompanying images. But he also acknowledges that his work will be received by an audience that is—to his mind—inherently biased and incapable of reading it properly. "There are several spirits in the world, some will take a thing one way, some another; a physician will not quarrel with a patient, because he refuseth to take the pill unless gilded, nor will we quarrel with any reader whose want of judgment or misapprehension misguides him to the finding of faults only, but pity him that his narrow capacities should so impotently desire that all others should be constituted after his size."[78]

The examples above all reflect the ways scientific authors attempted to influence a heterogeneous audience. We read their caveats and cautions both as rhetorical tools and as reflections of genuine underlying concern. Sometimes, however, the preface was written in an attempt to frame the reader's view of a text. William Gilbert, who in 1600 introduced the world to the first treatise on magnets and magnetism, asks his "Candid Reader, Studious of the Magnetic Philosophy" why he should even bother to publish, only to be "damned and torn to pieces by the maledictions of those who are either already sworn to the opinions of other men, or are foolish corruptors of good arts, learned idiots, grammatists, sophists, wranglers, and perverse little folk? But to you alone, true philosophizers, honest men, who seek knowledge not from books only but from things themselves, have I addressed these magnetical principles in this new sort of Philosophizing."[79] Gilbert's prefatory remarks accomplish two things: they justify the act of publishing, especially when so many books already exist, and they outline the kinds of readers he hoped to reach. Certainly, the individual holding Gilbert's book was not expected to put the text down after reading his preface, having self-identified as "perverse little folk." We cannot take Gilbert's preface at face value. We can, however, understand the way he uses his rhetoric of exclusion to prime his reader. The gentleman who stood in a bookshop and held his book would, somewhat like a good Calvinist, consider himself among "the elect" for whom the book was written and would presumably ascribe to himself those very characteristics Gilbert outlined. In the face of that preface he would be an honest man and a true philosopher. By prescribing the kind of reader he seeks, Gilbert effectively creates that reader, at least insofar as that individual views themselves. Gilbert has conditioned the reader's approach to his book.

Following Gilbert in this vein is Baruch Spinoza, the Dutch philosopher whose ideas were foundational to Enlightenment thought. Spinoza is a liminal figure in this study, straddling natural philosophy and ethics. His writings, like those of Descartes or Leibniz, reflect rationalist thought of the time, and his philosophy was influenced by Cartesian ideas about nature. For this study, Spinoza is valuable because, while he was an advocate for free speech, and in particular of liberal use of the printing press (without censure), he was unequivocal about his desire to limit the readership for his works. "It is thus plainer than the noonday sun," he writes, "that the real schismatics are those who condemn other men's books and subversively instigate the insolent mob against their authors." The authors themselves "for the most part write only for the learned reader and consider reason alone as their ally."[80] In the preface to his *Theologico-Political Treatise* of 1670, Spinoza explains his hopes for an audience that consists solely of philosophers, an audience he defines as "those capable of rational reasoning."[81] As for other readers, "I am not particularly eager to recommend this treatise to them, for I have no reason to expect that it could please them in any way." Spinoza goes on to describe the strength of people's prejudice and the power of their superstition, which for him includes religious devotion. Commoners who hold to such beliefs are, in his words, obstinate and impulsive.

> I do not therefore invite the common people and those who are afflicted with the same feelings as they are [i.e., who think theologically], to read these things. I would wish them to ignore the book altogether rather than make a nuisance of themselves by interpreting it perversely, as they do with everything, and while doing no good to themselves, harming others who would philosophize more freely were they able to surmount the obstacle of believing that reason should be subordinate to theology. I am confident that for this latter group of people this work will prove extremely useful.[82]

Responses to new ideas that were not grounded in the methods of natural philosophers both affected an idea's reception and created work for the author, who felt pressured to respond to spurious or poorly informed arguments. Yet we need not limit ourselves to the most renowned of scholars, the Newtons and the Descartes of the early modern world. Anatomist and surgeon

John Banister, in his brilliantly titled *The history of man, sucked from the sap of the most approved anatomists* (1579), addresses his prefatory letter to fellow surgeons:

> And I . . . from the depth of my heart renounce you, hoping assuredly, that from none of the flowers of this garden any of you shall take opportunity to suck that which may maintain the infection of your pestilent wretchedness hereafter. If therefore I have any where frequented a phrase above the common use of our English language, or bred words little different from the Latin, esteem the same to be done only for your cause, since . . . I have endeavored everywhere, to shade the kernel with a harder shell than you shall be able to crack. Away therefore you vipers. Let these my simple labors, whatsoever they are, be entertained in the hands of the true, virtuous, and honest artists and professors of Chirurgerie, that my expectation may be fulfilled, art rights advanced and God duly worshipped.[83]

Banister here echoes the posturing one sees frequently in these prefaces, summoning the very readers who would agree with him and rejecting in advance those readers who fail to do so. He also purports to encase his work in a metaphorical shell that is just hard enough to keep out those unfit to penetrate it. He writes for those he hopes will read him. Similarly, we can consider Dr. Thomas Sherley, physician-in-ordinary to Charles II and author of a treatise on kidney stones. In his preface he ironically expresses disdain for prefatory material, saying he would prefer to present his work "naked, and without an advocate [as philosophical subjects ought to do]." He would rather have his reader enter the book with no priming, no preconceived ideas about the material, so that they might draw their own conclusions. "This, I say, I would have done, could I have been assured, that this book should have fallen under the censure of none but philosophical, and knowing men, to whom I should have thought myself happy to submit my labours of this kind."[84] He claims to welcome the objections of learned colleagues and to embrace alternative theories if they are supported by reason and experiments. But he is wary of the unlearned reader, "a railing adversary fitter for my slight than my reply," such a reply not being worthy of his time. "I shall not fear censure, though I must be exposed to that of any man, which shall take the pains to peruse my book; I am not ignorant of the Proverb, *So many men, so many minds*: Nor of that other, *Habent sua fata libelli*: and therefore cannot expect that possibility of pleasing everybody."[85] The Latin phrase Sherley quotes is part of a longer saying attributed to the second-century Roman Terentianus Maurus: *Pro captu lectoris habent sua fata libelli*, "According to the capabilities of the reader, books have their destiny."[86] A fitting acknowledgment of the challenge any author of science or medicine faced.

Despite the many examples we have considered, not all prefaces were exclusionary in tone. To invoke Genette again, some vestibules were welcoming, inviting in anyone with interest and, at times, inverting the usual rhetoric by discouraging only experts. Numerous prefaces to works in medicine, alchemy, astronomy, and natural philosophy assured readers that the material they were about to encounter would not be prohibitively difficult, a comfort authors were willing to offer if such a readership served their needs. In his 1550 work, *A littel treatyse of astronomy very necessary for physyke and surgerye*, Anthony Ascham tells the reader of his preface that he has written the work "not for learned men but only for the unlearned English reader."[87] His aim in doing so, he claims, is to reveal to the layman God's omnipotent power as expressly seen in medicine and surgery. The contents are not technical, and Ascham has no desire to direct his work toward his peers. Rather, he uses it to establish himself as a capable and devout practitioner in the minds of the public. There is also Andrew Boorde's 1575 medical treatise, *The Breviarie of Health*. With four individual prefaces addressing physicians, surgeons, the sick, and readers, Boorde takes full advantage of the prefatory space as an opportunity to frame his ideas and outline his audience. In his letter to readers, he claims he has translated all difficult terms out of Latin, Arabic, and Greek, to facilitate reading, and has omitted certain medical details, lest "every bongler would practice physick."[88] And finally there is Thomas Brugis, whose preface to *The Marrow of Physick* promises readers access to secrets heretofore kept hidden by physicians: "Some men perhaps will thinke that nothing good or secret will be put in Print, because these kinde of bookes and very difficult to be published in English; others again knowing such things, would be loath to publish them and make the secrets of their science common, but I am rather of the Grecians minde . . . a good thing is the better the more common it is."[89]

Brugis claims to "eschew prolixity" so that "I would not willingly exceed the bounds of a preface making the porch bigger than the house." These examples are hardly outliers. In a survey of medical publications published between 1500 and 1700, 85 percent included a preface to the reader, and the majority echoed sentiments of Boorde and Ascham.[90] These prefaces are the closest thing we have to direct speech between the author and the individual seeking medical knowledge, and in many of them physicians preemptively defend themselves against critics, hoping to reinforce a measure of authority in the minds of their readers. They describe their experience and training, reference their medical influences, decry the censorious maneuvers of the college fellows, and—most importantly—articulate the democratic nature of their works. There would be no Greek or Latin, no veiled references to medicines only an elite few could concoct. Rather, the refrain of these prefaces is that medicine should be accessible to the vulgar.

Such prefaces were hardly unique to the medical field, nor were they dominated by authors. An English edition of Willem Janszoon Blaeu's *A Tutor to Astronomy and Geography,* translated from the Latin and printed by Joseph Moxon, includes Moxon's preface "The Publisher to the Reader." Here Moxon remarks on the demand for Blaeu's work among English-speaking astronomers, and he highlights areas where his translation differs from Blaeu's original language. "Some (but few) alterations I judged fitting to be made in this translation; and that because there falls out some difference between our Author's own globes, which he treats of, and the globes that I have given forth this last year."[91] Moxon extols the virtues of his globes for several pages, making it clear that his translation of Blaeu is, in fact, a vehicle for selling his globes and other instruments, but his marketing hardly undermines the intent of his work, which was to make Blaeu's instruction accessible "not only to deep wits, but to mean capacities." Moxon's appeal to the lay astronomer is not uncommon in prefaces penned by translators. Theirs was an egalitarian mission to make accessible works that would benefit the public, particularly in practical fields like navigation, accounting, midwifery, and medicine. The translator of Thomas Willis' *Practice of Physick,* who went by the pseudonym Eugenius Philatros, included a preface justifying the need for another English translation. It was, he said, a translation curated for common people, one that synopsized the most practical parts of Willis' tome for readers who needed access to the information, without the scaffolding of medical theory. Philatros also defends himself against charges that he merely copied someone else's English translation. "This is no transcript . . . I believe what I have here translated will be judged by all men to be more easy, correct, and clear, than what has been done before."[92]

Whether it be a preface from the translator, editor, or author, the prefatory space was a site of rhetorical positioning. This is true of prefaces across the subfields of early modern science, but it is clearly demonstrated in medical works. Nathaniel Lomax, a physician, tells his readers that he "aims to be as active in undeceiving the world, as empirics have been in deluding it."[93] In the preface to her book on domestic medicine, Hannah Woolley argues for the utility of her work, claiming that most other treatments of the topic were confounding rather than instructive.[94] And Thomas Law, in his preface to *Physick for the Poor*, explains that his work would make medical practice accessible to those of little means. He claims to offer "a complete method of physick, so plain and easy that the meanest capacities may attain it."[95] It is an injustice, he says, that those who cannot read Latin are deprived of important medical information, not to mention the inability of many to afford the expensive medicines that are recommended. But there is no better exemplar of a physician using print—and specifically the paratext of the preface—to his advantage than London practitioner William Salmon. A peripheral character in the history of medicine, Salmon is considered by some to be a middling physician, by others a mere quack. There is little contemporary information about him or his practice, save for what can be gleaned from his books. Indeed, his many prefaces to the reader are the only window we have into his life and experience, and while his prefaces employ the tropes we see elsewhere, they are the best source we have for understanding Salmon's practice. Studying them in combination with his actual texts, we begin to see in Salmon a paradigmatic example of the way authors used paratext to cultivate readers, to establish a reputation, and to expand a career.

William Salmon was born in 1644 and claimed to have learned medicine informally. As a young man he attached himself to an itinerant empirick who trafficked in homemade medicines.[96] This individual also traveled to New England, and Salmon accompanied him. By 1671 Salmon was back in London and had established a medical practice just outside the Smithfield Gate of St. Bartholomew's hospital. There he could treat individuals who had been denied care at the facility. Like many irregular physicians, he had a hand in all aspects of medicine: anatomy, chymical medicine, astrological physic, and even surgery. He practiced outside the purview of the college, unlicensed but certainly not unusual. In addition to seeing patients, Salmon developed, marketed, and sold his own line of medicines, claiming that his "Wonderful Pills" were "the most powerful Cathartick in the World." His "Cordial Pill," as he called it, was the "greatest and most excellent Preparation of all the Opiates yet invented." Coupled with his drops and balsam, Salmon claimed that he provided families with affordable and efficacious care. He even had a mail order business, offering to post medicines to anyone in the kingdom who sent for them.[97]

What makes William Salmon interesting among empiricks is his savvy use of the press to develop his practice. In a forty-three-year career, he published more than fifty treatises, which were sold at dozens of booksellers throughout London. He composed his publications carefully, drawing heavily on established authors and incorporating paratextual elements that, at the time, ascribed a certain authority to the works. His materials were organized and accessible, inviting a broad audience. Salmon was also strategic about which printers he worked with and where his books were sold, ensuring wide distribution in the London market. Through print, he erected a professional public image that defied accusations of quackery.

Salmon's first publication, *Synopsis Medicinæ, a Compendium of Astrological, Galenical and Chymical Physick in three volumes*, appeared in 1671—the year he opened his medical practice. It was printed in octavo by William Godbid, a printer known for producing musical and mathematical works, and was sold for 3s 7d at the shop of Richard Jones, bookseller at the Golden Lion in Little Britain. We can get a sense of the atmosphere in this part of London from a contemporary observer, who described it as "a plentiful and perpetual emporium of learned authors . . . and the learned gladly resorted to them, where they seldom failed to meet with agreeable conversation; and the booksellers themselves were knowing and conversible men."[98] Salmon had chosen a printing hot spot for his initial publication, and his work was well priced; at under four shillings, the *Synopsis Medicinae* was slightly cheaper than an evening at Blackfriars to enjoy the theater.[99]

Diverse in content and prolix in presentation—it was 432 pages—the *Synopsis* covered tremendous medical ground. Galenic ideas were woven into new, chymical theories of medicine, offering readers, if not an internally consistent medical theory, an eclectic one. In his prefatory epistle Salmon touts the work's accessibility, arguing that obscurity only hinders public health, and he describes the three books that constitute the whole: diagnostics, prognostics, and therapeutics. For Salmon each had both a physical and astral component. He was also keen to preemptively dismiss any accusations that his work was unoriginal, merely recycled theory. Salmon maintains that nobody had yet braided the threads of chymical, herbal, uranical, and classical medicine together in such an accessible way. He hopes "that although [I] may not be able to answer the expectations of all nor make [myself] become very useful to the

Learned, yet [I] may sufficiently satisfy the curiosity of the ignorant."[100] We do not know precisely how many copies of the *Synopsis* sold, but a second, expanded edition of more than 1,200 pages appeared in 1681, and that edition was reissued in 1685, 1695, and 1699. William Salmon's first foray into print was a success.

Once he had published on medical theory and praxis, Salmon turned his attention to pharmacology, a move that put him in tangential conflict with the College of Physicians. In 1618 the college published, in Latin, an official inventory of medicines, entitled *Pharmacopoeia Londonensis*. Replete with the most commonly prescribed medicines, this was the standard formulary in medical practice. It was also carefully guarded by the college, which sought to withhold such knowledge from commoners and thereby preserve the prescribing authority of physicians and recognized apothecaries. In 1649, however, herbalist and anti-establishment physician Nicholas Culpeper published an English translation, cracking open nearly two thousand medical recipes for the public to access and vigorously promoting the accuracy of his text. He also got a jump on piracy—which he himself would be accused of—by assailing any "thief" who would dare to steal his work.[101] For Culpeper, licensed physicians were among those thieves; he therefore viewed his translation as an effort to wrest control away from the college fellows, rather than a piratical one that usurped legitimate knowledge. It was in this same spirit of democratizing medicine that Salmon, in 1678, offered laymen another choice with his English translation of the *Pharmacopoeia Londonensis*. Subtitled *The New London Dispensatory*, it was brazenly dedicated to the king, and—Salmon says in his preface— translated for the public good. The work offers "a translation of the London Dispensatory, lately reformed by the fellows now living . . . To which we have added certain animadversions upon their preparation."[102] The publication was an overt challenge the college's authority to regulate drugs and apothecaries, and it was only the second English translation made available in London. Salmon's translation targeted lay readers with its clarity, portability (it was produced in octavo), and affordability. Its content was also expanded beyond what the college offered, with preparations described in both traditional and Paracelsian terms, and Salmon claimed to offer the public remedies that would otherwise be kept hidden by the college. In the preface to the reader, he directly criticizes the college for keeping its *Pharmacopoeia* out of the reach of those who needed it, but he did not limit his criticism to the elites. Toward the end of his preface—with the added highlight of a manicule—he takes on an "impudent and illiterate quack-salver" who had gone around London purporting to treat people but caused harm instead. This person claimed to be "William Salmon," an identity fraud that left the real Dr. Salmon undone. Salmon chastised the imposter, as well as all misguided charlatans, and adamantly argued for his place among legitimate, "honest" physicians.

The *Pharmacopoeia* was printed by Thomas Dawks and was available at booksellers across London. These included Thomas Bassett, John Wright, and Richard Chiswell, who all advertised the work. Bassett's shop was reputable; he was known for his catalog of law and history and also for having published the work of anatomist Thomas Willis.[103] He offered Salmon's *Pharmacopoeia* bound for 7 shillings. Richard Chiswell was even better known, having published more than one thousand books in the 17th century.[104] His father-in-law had been bookseller to the king, and Chiswell himself became a printer to the Royal Society and was considered one of the great scientific publishers of the day.[105] Salmon had therefore taken one of the most important medical works of the period, the *Pharmacopoeia*, translated it into English and had it sold in the most popular bookstores in London for buying scientific and medical publications. It would reappear in seven editions over Salmon's career, continually printed by Dawks but sold at an increasing number of shops. Salmon used print to stake out ground for his practice.

Salmon's next major foray into medical publishing involved his *Iatrica: The practice of curing*. It also marked his most creative publication strategy. Described on the title page as a work "of singular use to all the practicers of the art of physick, and chirurgery, whether physicians, chyrurgians, apothecaries, or charitable as well disposed gentlemen and ladies," the *Iatrica* was to be a massive compendium of medicines. Salmon estimates in his preface to the reader that the entire thing would constitute more than five hundred pages and take years to complete. Given the size and scope, he decided to release pages of his *Iatrica* as they came off the press: "For the present satisfaction of the buyer . . . a continual supply [will] be provided to the carrying on of this so great a work, which, were it to be exposed whole, would scarcely be sold under 4£ a book. The first sheet will begin Wednesday the 27th of July 1681, and so come out weekly, Wednesdays and Fridays, till the sum of all the said cures are exposed."[106] Salmon notes that each sheet released would treat a particular disease and that sheets could be collected into groups forming five books. The first section deals with diseases of the head, and the work would conclude, four hundred pages later, with a theory of apoplexy. It was originally printed in quarto for Thomas Dawkes (the younger), printer to the king, who had shops throughout London, and for Langley Curtiss, a bookseller on Ludgate Hill. Copies were sold by Thomas Passinger, who also had available in his shop Dr. Salmon's medicines.

Salmon followed *Iatrica* with *The Family Dictionary* (1695), an alphabetical list of domestic medicines and cures.[107] In his preface to the *Dictionary* Salmon writes, "Here you may repair to an asylum and find the arcana opened for your good, which have been hitherto locked up from the greater part of mankind. All difficulties and hard terms, or words, have been removed, which have puzzled ingenious persons, and the path is made so smooth that any reasonable capacity or understanding may travel in it without the least fear of stumbling or falling into error."[108]

Thus far we have seen William Salmon publish a general and accessible work of physick, a comprehensive pharmacopoeia, a more specialized work on treating diseases, and a layperson's dictionary of medical terms and recipes. In the midst of this he was also producing annual almanacks, broadsides, and pamphlets. By the mid-1680s there were at least four major booksellers in London who offered all of Salmon's works for sale, and a dozen more had at least one. The only notable omissions to this corpus at that point were works on surgery and anatomy, but the former he addresses in 1687 with *Paratērēmata; or select physical and chyrurgical observations*. Here Salmon summarizes different surgical practices as they related to particular ailments. Engraved plates show the reader which instruments to use. The latter he treats in *Ars Anatomica*, a work developed over many years but only published posthumously in 1714, whereupon it appeared in at least five bookshops. Salmon was no anatomist, so this work was a compilation of other treatises, but by the time it appeared his reputation was sufficiently established that the book would have sold.

The next two decades saw Salmon issue more than a dozen works from familiar and new publishers. He even had a hand in medical translation. In 1686 he translated Johann Dolaus's *Systema medicinale* into a "portable volume for the public good," and in 1694 he translated the work of the Dutch physician and anatomist Isbrand van Diemerbroeck—*The Anatomy of Human Bodies*. His most popular translation, though, was of Thomas Sydenham's *Processus Integri*, or *The Practice of Physick*. The book was published in 1695 by Sam Smith and Benjamin Walford, printers to the Royal Society. Salmon's translation included considerable annotation and editing, making Sydenham's medical ideas widely accessible.

PRAXIS MEDICA.

THE
Practice of Physick
O R,

Dr. SYDENHAM's *Proceſſus Integri*, Tranſlated out of *Latin* into *Engliſh*, with large *Annotations*, *Animadverſions* and *Practical Obſervations* on the ſame.

CONTAINING
The *Names*, *Places*, *Signs*, *Cauſes*, *Progno-noſticks*, and *Cures*, of all the moſt Uſual and Popular Diſeaſes afflicting the Bodies of Human Kind, according to the moſt approved Modes of Practice.

Among which you have

The Pathology, and Various Methods of Curing *A CLAP*, or *Virulent Running of the Reins*, and the French *POX*, with all their *Attendent Symptoms*, beyond whatever was yet publiſh'd on this Subject by any other Author, Ancient or Modern, ſince the Diſeaſe firſt appeared in the World to this Day.

The Third Edition, Inlarged throughout, with ſome Thouſands of Additions not in the firſt Impreſſion.

By WILLIAM SALMON, M. D.

LONDON,

Printed for *J. Knapton* at the *Crown*, and *W. Innys*, at the *Prince's-Arms* in St. *Paul's* Church-Yard, 1716.

William Salmon, ed., *Dr. Sydenham's Practice of Physick* (1695). *Courtesy of the National Library of Medicine.*

That Salmon found a kindred spirit in this work is not surprising. Sydenham was widely lauded as the English Hippocrates, and at times he operated outside of the medical mainstream. He had a medical practice for thirteen years before he even received his MD, and although he was licensed by the College of Physicians in 1663, he never applied to the fellowship. His calls for a reform of medicine, his belief that medical interventions were not always good or necessary, and his dedication to empirical learning, all had echoes in Salmon's work. The translation of Sydenham was, for Salmon, a project clearly aligned with his philosophical interests. But Salmon was just as aware of the book's commercial potential. Signaling his attentiveness to marketability, he noted that the Latin edition had sold more than seven thousand copies in three impressions, so a translation would sell easily. In fact it did, going through three editions. But it also aligned Salmon in a new way with mainstream medicine in London. College physicians thought very highly of Sydenham; if Salmon could position himself in proximity to the new Hippocrates, he stood only to benefit.

Salmon's most direct engagement with the politics of medical licensure—and his boldest use of print—came in 1698, when he published *A Rebuke to the Authors of A Blew-Book; call'd "The State of Physick in London," which is indeed the black and blue state of physick*. With this work Salmon entered what is known as the "dispensary controversy." At its root was a debate between the College of Physicians and the Apothecaries Guild, which is why Salmon's title page indicated that he wrote the work on behalf of the apothecaries and surgeons of London.[109] By printing this, Salmon entered a public quarrel with the College of Physicians that had major implications for his status as a legitimate practitioner.

Salmon's rebuke was a response to the Royal College of Physicians' publication, earlier in the year, entitled *The State of Physick in London*.[110] The lengthy subtitle for this pamphlet referenced the college's plan to start offering free medical advice for the poor and to increase their role in preparing medicines. The work begins with a brief history of the college and the acts of Parliament that gave it authority to punish illegal practitioners. It then describes the formation of the Apothecaries Guild in 1617, highlighting the problems that subsequently emerged. Among those were the sheer increase in the number of apothecaries in London (over eight hundred according to the treatise) and the exorbitant fees they charged for medicines. London's surgeons fare little better, coming under fire for abuses and fatal mistakes in practice. After a dozen pages disparaging the ethics of these practitioners, the college pledges to publish a list of "cheats and imposters, who swarm about the kingdom, to the visible destruction of thousands of our fellow subjects."[111] The pamphlet came from the press of Elizabeth Whitlock (fl. 1695–99), who assumed responsibility for her husband's press upon his death. Her shop was near Stationers Hall, and she published everything from political to religious tracts, as well as a few notable works in defense of the barber surgeons.

William Salmon's response, *A Rebuke to the Authors of A Blew-Book*, came swiftly. And, to heighten the tension, he had it printed by the same Whitlock who had published the college's piece. Salmon's rhetoric typifies the pamphlet wars of the day. He calls the college's treatise, "a Fardle or Bundle of Lies, containing scarcely one grain of honesty."[112] He then takes up their claims point by point, quoting their text in italic and then responding. At one point he examines the college's proposal that they make medicines available to the poor at their "intrinsic value," by which they meant at cost, without profit. Salmon scoffs at such a plan: "Here's a great cry, and a little wool, as the devil said when he shore his hogs."[113] Pivoting to his own publication, he mentions his *Pharmacopoeia*, highlighting the accessibility of pharmaceuticals central to all his works. Ultimately Salmon makes the rhetorical shift that underscores his entire corpus: that the college itself is a confederacy of charlatans. "I do not think any quack-bill that was ever published, can equal this of the Blew-Book-men; it is perfectly Quacking all over, with all the hyperboles that the thing will admit of."[114] He calls their work typical Bedlam discourse and stresses that the public cares very little about the college's magnanimous posturing when they cannot do something as simple as access a physician. His polemic is anti-monopoly, anti-corporate, and decidedly populist. Whether this thirty-two-page pamphlet, printed in octavo, sold like mad or gathered dust, we cannot know. It was for sale at just one shop, and, given the absence of paratext that was typical for Salmon's publications, its audience is clear: he was speaking to the college physicians directly, reframing them as the quacks the public should distrust, and he was speaking to his established readership, reminding them that his work on their behalf—though unlicensed—was valuable.

This episode was just one of many quarrels between the college and unlicensed physicians, but it represents the degree to which Salmon—through print—had come to symbolize a real threat to the college. Near the end of his life, Salmon issued a second edition of his translation of Sydenham. In his preface he recounts all he had brought to the public, highlighting sixteen of his works that he judged most important. The *Botanologia* was first, *The Art of Anatomy* last. Salmon was reflective, considering not only the nature and use of his published works but the medical hierarchy generally.

> There are a sort of men in the world who affect to be Doctors, but are no Physicians, who tho' they are great pretenders to our Art . . . know so little truly of Physick . . . To such Persons as these we address not this work. These, by a kind of Pedantry, triumph in their ignorance and Folly . . . to the Consideration of such this work is not commended, because we know it is as much above their understandings, as the Author thereof is below their thoughts.

He closes by suggesting that all the books he has published, while publicly condemned by these learned physicians, have been secretly read by them, proving at once the pedagogic value of the books and the hypocrisy of the learned men who pretend to hate them. For my purposes, Salmon is paradigmatic of the way print could be used in early modern medicine to establish not just legitimacy but authority. Like the herbal and medical texts of Nicholas Culpeper before him, Salmon's treatises found their way into the libraries of physicians and the bookshelves of households throughout England and the North American colonies.[115] By publishing works on a variety of medical topics and making those books accessible at dozens of shops throughout the city, Salmon shifted the lines that demarcated quackery, pointing his manicule at both the top and bottom of the medical hierarchy, while planting his feet firmly in the middle.

Salmon's career was the embodiment of the populist tone found in many medical prefaces. The authors of such prefatory appeals eschewed complexity and scholarly trappings and instead promoted a practical and accessible physick. They also demonstrated the ways irregular physicians began to use print to shift the definition of charlatanism. Rather than defending themselves against the characterization that they were the quacks or denying that empiricks were harmful, the irregulars openly complained about the problem. Everywhere, they said, people were being treated poorly by individuals who did not know what they were doing. To make matters worse, these imposters to medicine were not necessarily street physicians or pill hawkers but sometimes educated and licensed practitioners who—despite the collegiate stamp of approval—knew little of what they spoke. The irregulars sought to redefine medical competence and used print to do it. And so a 1675 treatise by Thomas Cooke warns readers of two equal dangers: charlatans selling unsafe cures, and the college creating a market for those cures because licensed help was not available. John Colbatch, an apothecary, tells his readers that although he expects "carpers" (a term used to signal collegiate attacks), his design is to *undeceive* people. He addresses the rift between college physicians and apothecaries that, he says, needs mending lest quacks take advantage. "Common people who are not judges are easily imposed on."[116] Abraham Miles likewise begins his popular work of physick by assailing those learned physicians who keep their knowledge locked up, available only to themselves. *They* are the mountebanks, not the doctors who publish accessible cures. Another physician assures his reader that his desire is "not [to] break down your banks, by letting in a flood of Empiricks upon you," but to offer recourse to those who cannot access a licensed practitioner. Treatise after treatise warns against both the reckless empiricks and the Royal College.

Such solicitous prefaces were common from medical authors, but they hardly had a monopoly of this kind of prefatory posturing. Scholars of natural history, botany, husbandry, and the domestic sciences also published works that invited the reader in, first through a preface, and then on to the studies themselves. An author who used his prefatory epistle to great effect is Hugh Plat (1552–1608). In Plat we see interesting ideas about print from a London resident who, though nominally a lawyer, was deeply connected to the applied sciences of day-to-day life: brewing, gardening, domestic chemistry (preserving, canning), medicine, surgery, and assorted technologies such as engines and pumps.[117] Plat was keenly interested in the natural sciences and cultivated relationships with other learned men of his time. Deborah Harkness' study of Plat, wherein she examines his career alongside the better-known Francis Bacon, is foundational to my analysis. She shows how Plat's career exemplified the Baconian program for producing knowledge of nature and how, in both his methods and his wide array of interests, Plat was a quintessential gentleman of early modern science, despite the fact that his name fails to rank among the notables of the period. In drawing Plat into my own study, I aim to show how publishing strategies allowed Plat to achieve many of his ambitious goals in the sciences. His first printed work, *The Floures of Philosophy*, appeared in 1572, but he is best known for *The jewell house of art and nature*, an extensive compendium of practical knowledge—from the improvement of soil to the brewing of beer—intended for use in everyday life.

Plat begins *The jewell house* with a standard dedication to Robert, Earl of Essex, at once self-effacing and self-aggrandizing. The secrets of nature, "long since interred in a case of marble," he would "break out of their tombs." Plat positions himself as a risk-taker, one who dares to present his ideas despite the potential attacks, "not only of the base and vulgar sort of people . . . but also of those that have had some better education, and been brought up in the same schools, yea under the same tutors with them." His characterization of his readership is somewhat disingenuous, given that a great deal of his information and experimental data came from commoners around London, with whom Plat interacted regularly. But the prefatory rhetoric aims to insulate him from criticism, regardless of what quarter it came from.

It is in Plat's preface to the reader, however, that his posture becomes clearer. He has no time for those who would volley uneducated criticism. They are, he says, "scarcely scavengers in the ward . . . understanding no more Latin than *Noverint universi*."[118] The tone here is one of a gentleman refusing to engage in fruitless debate with a layman, but Plat goes on to characterize himself as "a young novesse in the schools of philosophy . . . a slender husbandman in the rights and culture of the ground." A scholar of soil, he accepts, might not be seen as someone keen on contributing to the world, but Plat defends his pedigree, or at least his work, suggesting that God is not bound to any particular age or profession or estate. Plat's God is more egalitarian than that, and his book, presenting the insights God gave him over the course of many years' experience, will prove useful to thousands, or so he hopes.

Plat returns to his experience time and again. In the preface to his *Floraes Paradise*, a work devoted to gardening, he offers his "studious and well-affected reader" a strong case for empirical knowledge, validating his own authority on gardening based on "particular experience . . . obtained by a pretty volume of experimental observations."[119] He has, he assures his readers, spoken of these ideas with gentlemen of good skill, so we have reason to trust his findings. But Plat does more than legitimize his findings based on experience and experiment. He claims to also present his work in a practical way, fit for manual workers. Not so subtly, he narrows the description of his intended audience, "the workmen" who are in the soil, not the philosophers who ponder husbandry from their desks. "I leave method at this time to schoolmen . . . whose labors have greatly furnished our studies and libraries, but little or nothing altered or graced our gardens and orchards." Plat's is a call to empirical practice, geared toward "those who seek out the practical, and the operative part of nature, whereunto but a few in many ages have attained."[120] His knowledge, accordingly, is not "an imaginary conceit in a scholar's private study, but wrung out of the earth by the painful hand of experience." He effectively turns elitism on its head, rejecting the academic or philosophical in favor of the pragmatic and empirical. Plat embraced print because it was through publication that he could achieve the kind of standing and renown he sought.

Plat's example is useful in understanding the subset of prefatory epistles that are welcoming. For many authors in the sciences, publishing was closely tethered to issues of authority. I suggested at the outset of this chapter that a broad readership and widespread circulation often threatened to dilute the authority of the author. The chorus of complaints about the existence of "too many books" reflects this concern. For scientists at the top of their fields—for Kepler or Huygens—print offered little that they could not achieve through manuscript circulation. Communication networks were vast in the early modern period. Twenty thousand letters were exchanged among scholars in the Dutch Republic during the seventeenth century, and Henry Oldenburg alone was involved in the exchange of over three thousand.[121] But print allowed authors with less fame, or with fewer links to a patronage system, to accrue authority through their publications. It also helped them make a living in a way that more elite authors did not require. Hugh Plat was well-educated aristocrat, but he pitched his treatises to "the workmen" and consciously disavowed the scholars who, he claimed, produced nothing more than words. For Plat, printing was not a liability but an opportunity not only to share his ideas but to augment his own reputation. By publishing books, pamphlets, even broadsides, all of which heralded his empirically gained knowledge, Plat established a level of authority he would not have attained through manuscript circulation.

For scientific authors it was an exciting and fruitful time. It was also an uneasy time, when new ideas about the universe or the body could bore holes into centuries of accepted wisdom. In other chapters I examine the various ways scientific authors attempted to circumvent the public nature of print culture. Their collective reluctance, I believe, is best illustrated in the paratext of reader's prefaces. Many of the authors discussed here, and numerous others not included, used the preface to erect a wall, hoping to keep some readers out while inviting other inside. Historically, prefatory expressions of reluctance to publish have been ascribed to authorial posturing. A scientific author who purported to be disinterested in publication gained credibility. But such an interpretation does not fully capture what authors—especially in the sciences—sought to accomplish when they included a preface. Astronomers, mathematicians, and natural philosophers may have expressed a desire to retain the dignity of the gentleman-scholar, but they also genuinely hoped to use prefaces to the reader to influence the reception of their work. In some cases that meant building a vestibule that was accessible and welcoming, while in other instances it meant an inhospitable or challenging antechamber, built to inoculate the author against the tiresome retorts of the unskilled. Randall Ingram notes that prefaces "record emerging assumptions about the production, circulation, and proper use of texts."[122] In that sense they act as archives not of reader responses or authorial intention but of authorial postures. In early modern science, authors frequently exercised their rhetorical skills in these prefaces, but if we attend to them as paratext that informs a work's reading, we gain insight into what scientific authors hoped to achieve—and avoid—when their books were read.

Chapter Three

The Controlled Distribution
of Scientific Works

> You are, Sir, among the few men for whom this book was made, and who I
> hope to have as readers; because of this I am obliged to send you this
> example.
> —CHRISTIAAN HUYGENS TO PIERRE BAYLE, FEBRUARY 17, 1690

As we have seen in preceding chapters, scientific authors acknowledged the advantages of the printing press. By the seventeenth century, they had grown more familiar, and adept, at negotiating the world of print. To effectively communicate their ideas to the appropriate audiences, authors had to consider how they wanted to shape their publications, to whom they should dedicate them, how many copies to print, and how best to circulate them. In these and many other respects the authors of scientific publications operated in three worlds. Foremost was the world of theories and ideas; another, the humdrum world of typography and print runs; and finally, the ever-changing world of politics, diplomacy, and social status. This chapter focuses specifically on the circulation of texts and the ways scientific authors took advantage of distribution networks to carve out ideal audiences for their work.

It was not uncommon in the sixteenth and seventeenth centuries for an author to control—at least to some extent—the circulation of their publications. Tycho Brahe, René Descartes, Galileo Galilei, and Johannes Kepler are just a few of the scientists whose texts were carefully distributed to chosen recipients, though the control over distribution varied widely from one case to the next, as did their particular motivations.

A common reason for sending copies of one's books to chosen recipients was to cultivate patronage and forge new relationships with other scholars. Historians have examined this in the context of gift giving, especially in early modern patron-client relationships.[1] Implicit in each exchange was a transfer of power and credit that could potentially afford the author higher status, monetary rewards, and recognition in a given field. For a mathematician like Galileo, this meant a new position as court mathematician for the Medici, a notable step up from his teaching position in Pisa. For Christiaan Huygens, who was already wealthy enough to carry out his studies as it pleased him, earning the patronage of Louis XIV meant a position in the prestigious Académie Royale des Sciences in Paris. But even more modest relationships could be useful, offering scientists entry into tightly knit groups of scholars who might be willing to share ideas, if not manuscripts. In a period when science was often as competitive as it was collaborative, it was a distinct advantage to control the destination of one's published work.

Another motivation behind targeted distribution of one's work was the desire to secure priority. Judicious selection of an audience could assure a scientist that credit would be properly assigned. So pressing was the issue that authors like Huygens and Kepler spoke often of the secrets they would soon publish, theories Kepler delightfully describes as "not yet fully grown and half-fledged doves."[2] In some cases, these secrets were published *as secrets*, through coded messages such as anagrams whose solutions would only be revealed when the author was ready. Galileo's announcement in a letter to Kepler that he believed "smaismrmilmepoetaleumibunenugttauiras" hardly created shockwaves. Head-scratches may be more like it, though Kepler still attempted—albeit unsuccessfully—to crack the code.[3] And Huygens published a pamphlet in 1656, discussed below, which announced a new theory in anagram form. What was privately known remained so, despite being printed and circulated. And because the idea of intellectual property was not codified in this period, scientific authors found that managing their audience through controlled circulation of texts was worth the effort. If they did nothing and left distribution up to the chance purchase from a bookseller's stall, they missed important, even critical, opportunities to gain the credit or attention they sought.

Finally, personally described networks allowed authors to send their work to designated readers who would, it was hoped, ensure a fair reading. The previous chapter touched on this by exploring the prescriptive rhetoric of prefaces to the reader. When Kepler referred in his preface to the "idiots" who would be wise to put his book down, he revealed the uneasiness scientists felt regarding wide exposure to a general audience. Putting a copy of one's ideas in another scholar's hands, sometimes with an inscription and other times with a request for feedback, intrinsically gave the author a measure of control.

Though the printing press had flourished for more than a century, and publication was important, personal relationships continued to be essential within the scientific community. Gifting one's book to a fellow scholar fostered these relationships.

The majority of scientific authors disseminated copies they received from the printer as payment for the manuscript, since few printers paid money for the right to publish a work. When an author turned over their original manuscript, or "fair copy," they did not receive payment in exchange but rather a number of complimentary copies upon completion of the run.[4] John Ward, a diarist and physician, notes in his journal that "in printing books, this method for the copies in the first impression; they give the author 200 copies at half the price that they may bee sure to have some taken of, the 2d: edition they give him intirely one in ten."[5]

This number could vary widely, depending on the status of the author and the size of the run. John Evelyn, frustrated with bookseller Richard Chiswell, attacks such parsimony in a letter: "I did not believe you would have been so backward in gratifying me with a few copies of my own book, designed for some friends of mine to whom I could not (in good manners) omit the presenting them with a trifle they so long expected . . . and I have the more reason to complain of it, because . . . I might have had 100 copies more (with many thanks) from another booke-seller, whilst I was working for you gratis."[6]

Conversely, René Descartes' contract with Jan Maire for the *Meditations* stipulated that three thousand copies would be printed, a large run for the period.[7] As payment Descartes received two hundred copies for his own distribution, though Maire would retain the manuscript.[8] In other cases they were purchased outright by the author, as when Johannes Kepler purchased two hundred copies of his *Mysterium Cosmographicum* for his personal distribution.[9] While many authors received such copies, the best exemplar of authorial distribution among scientists is arguably Christiaan Huygens. Without fail, he distributed copies of every work he wrote, an average of fifty books per publication, and he consistently recorded who would receive these presentation copies. Found among his papers—on the back of a letter or across from some doodled sketches—these lists are all structured similarly, with the most important recipients written down first and other individuals following. The names appear in groups, often according to nationality. Huygens also organized subgroups according to the recipients' disciplinary affiliation. Not only do these handwritten lists provide a unique view of Huygens' targeted audiences; the organizing principles he used in assembling them provides insight into how he viewed his own readership in social, intellectual, and sometimes political terms. We see through Huygens' publishing history the entire array of motivations that prompted scientists to circulate copies of their works. If they were going to print, they would exercise their ability to guide their works into just the right hands.

There was great variability in how a scientist would want a work printed, what the treatise would look like coming off the press, and the number of copies they would have for personal distribution. At one end of the spectrum was the Danish astronomer Tycho Brahe, who controlled the printing of his works on his private press, a privilege I will explore further in chapter 5. With no financial constraints, he could afford sizable print runs, often on fine paper, with high-quality ink, and nicely bound. Some copies were made commercially available, at the Frankfurt book fair, for example, but generally Brahe saw to the circulation of the work. Consequently, his chosen audience was potentially *the only* audience for a work. He gave copies of his 1598 *Astronomiae Instauratae Mechanica*, of which only one hundred were printed, to the courts of Rudolf II and Princess Christina of Sweden.[10] Other copies were sent as gifts to dukes and princes and professors in The Hague, Munich, Copenhagen, Florence, and Prague. He used his books in multiple ways, cultivating different audiences for different purposes, but this was only possible because he circumvented traditional printing and publishing.[11] Few scientists, however, shared Tycho's level of control. Exceptions include Johannes Müller, mentioned in the introduction, and Johannes Hevelius of Danzig, both of whom directed the distribution of their works. Overall, however, we should view these men as exceptional. They enjoyed resources unavailable to most scientists.

More common were the authors who had little or no control over their works' distribution once they were published. Either they lacked the wealth and social standing to select exclusive readers, or they were beholden to a publishing system that required broader marketing and sales. There were also authors whose work was distributed in a targeted fashion but by someone other than themselves. For example, patrons—rather than authors—served as the nodal points for wider circulation of a text. They might send copies to other patrons, aristocrats, or learned societies, as Prince Leopold did with Borelli's *Theoricae Mediceorum Planetarum* and the Cimento Academy's *Saggi di naturali esperienze*.[12] Galileo's *Sidereus Nuncius* was circulated by his Medici patrons, and Kepler's *Astronomia Nova* was given out only with permission by

Rudolf II.[13] For the client in these exchanges—the scientific author—this could be advantageous since the prestige of the patron would undoubtedly enhance the prestige of the work. Moreover, if the work was controversial, distribution by the patron could quell potential disputes or even circumvent censorship.[14] The patron also assigned prestige to a work that a scholar might not otherwise have. Tycho comments on this very case in his *Astronomiae Instauratae* when he explains why he sent works to a very select audience: "The reason [I limit circulation] is that I do not want such precious inventions to lose their value by being known to everybody, as often happens . . . To distinguished and princely persons, however, who might be especially interested in such matters, and likely to cultivate these sublime sciences with greater ardour than others and to further them in a liberal and laudable way, only to such persons shall I be willing to reveal and explain these matters when convinced of their gracious benevolence."[15] We can look to a number of scientific authors who distributed their works in strategic ways, but none engaged in the practice with as much forethought and consistency as Christiaan Huygens. To track Huygens' dedicatory copies is to engage in a broad-ranging study of the motivations behind, and execution of, an author's circulation of their work.

Huygens, a Dutchman of wealth and social standing, established his reputation in mathematics as a young man. His education with tutors was supplemented by correspondence with other mathematicians, including the Parisian philosopher Marin Mersenne. Though Huygens originally reached out to Mersenne through his father, Constantijn, who acted as a conduit for reasons of politeness and protocol, Mersenne soon recognized the talent of the young man and allowed direct correspondence.[16] He was an important link for Huygens because he sat squarely in the center of a prolific circle of mathematicians, coordinating exchanges of demonstrations and managing various dialogues between authors. Word of Huygens' aptitude spread quickly. In a letter to René Descartes his father called him "my Archimedes," and by his twenty-second birthday his talents were on display in print. *Theoremata de Quadratura* was published in 1651 by the Elzeviers in Leiden. In it Huygens treated theorems related to the area under hyperbolas, ellipses, and circles, among other topics of mathematical analysis. Three years later Huygens published *De Circuli Magnitudine Inventa*, a display of his geometric skills in determining a more accurate value for pi. Problems centering on the area of the circle and estimations for pi were central to mathematical pursuits of the period, and

Huygens' contributions in this area were significant. Appended to the 1654 work was a separate treatise intended to display his geometrical acumen. It contains solutions to a number of classical problems that were simpler or more elegant than those previously proposed.

Drawing on the social and diplomatic connections his father had to *savants* in France, as well as his own newly formed correspondences, Huygens thoughtfully distributed his mathematical publications to individuals who could follow his methodology and support his results. Doing this allowed him to solidify an important network of supporters and gain a valuable reputation among mathematicians. Additionally, a close look at changes in the recipients from 1651 to 1654 shows the extent to which Huygens' community of scholars expanded. He chose problems in geometry that were vigorously debated at the time, and his distribution of books helped him nurture alliances that would support his approach in these contested areas, particularly where the quadrature of the circle was concerned.

Huygens first circulated the *Theoremata de Quadratura*, a forty-three-page work printed in quarto, in December 1651. According to a list of names on the last page of a correspondence notebook, Huygens sent out thirty-four copies of *Theoremata* over the course of two years. The names have notes next to them indicating the date that a copy (or copies) of the book had been sent. On the first line we read, "26 Dec. wrote to P. Greg. à S. Vincentio with my Theoremata. One to P. Sarasa," the "P" designating *pater*, or "father," for members of the Jesuit order.[17] Two more copies were sent before the year's end, both to Leiden mathematicians. Distribution continued in 1652 with the note for January 24 indicating, "to P. Seghers with 5 examples for la Faille. Claudius Ricardus, P. Tacquet. 2 for P. Greg." The task of delivering at least nine copies was delegated to the Flemish artist Daniel Seghers, a Jesuit known for his floral still-life paintings. Because Seghers traveled widely and had a broad network of correspondents, from Jesuits to aristocratic patrons, he was an ideal person to manage communication between *savants* and members of his society, as Huygens' letter indicates:

> I find myself obliged to personally thank you for your attention and care in maintaining correspondence between [Jesuit mathematician] Father Gregory and myself. Knowing that you were capable of distributing my books, which is to my advantage, I would have sent you the enclosed copies of my books at an earlier time, but the wintry conditions prevented me from doing so. I cannot

be more pleased than when I can inform my learned colleagues of my inventions. I will consider myself fortunate when, through you, each of these booklets could be distributed to the established addresses.[18]

All the recipients reached through Seghers were Jesuits, linked through their educations and their collaboration on mathematical projects. Huygens understood that by generating a dialogue with them, he would establish himself within the network that was most active in mathematics at the time. As he wrote to his brother Lodewijck in early 1652, "I have begun to be known among these Reverend Fathers, and it will really be to my advantage when Father Seghers has distributed the examples which he asked for . . . to knowing men of that company [Jesuits] and their acquaintances."[19] He was right, too: to gain favor among one or two of this order was critical. As Descartes once wrote, "The Jesuits are in close correspondence with each other; so the testimony of just one of them is enough to make me look for them all to be on my side."[20] With his 1651 publication, Huygens used mathematics and presentation copies to open a new door.

Huygens' next publication was also mathematical, and again he distributed copies to strategically chosen recipients. A letter to his colleague announces, "I am sending you a copy of my recently published book, Great van Schooten, and none do I send more willingly, than to you, whom I naturally esteem above all others, the most skilled and benevolent of our peers."[21] This letter is found among Huygens' papers in Leiden, and on the back of it he wrote a list—next to some random calculations and a sketch of a flame—of the people who were to receive copies of the book: "Gool, Schoten, le Ducq, Kraen, vander Wal, Chanut, C. de Briene, Blondel, Kinner, Vincent, Sarasa, Bibliot., Tacquet, Gutsch., Ick, Stevin."[22] Meanwhile, Huygens's father was also sending out copies. The Palatine princess Elisabeth and the diplomat Johann Jacob Stöckar both received their copies of Huygens' book from Constantijn. He encouraged Stöckar to consider the work in his studies, assuring him that he would find the material worthy. "This young man is preparing other works for the press, among which those concerning the telescope will be most worthwhile."[23] In December of that year, Constantijn also writes Princess Elizabeth, "My Archimedes has again brought out a certain new production of his genius; I hasten to prevent the reproach with which Your Highness honored me in the past, when I was late in giving you an account of him from the beginning."[24] The elder Huygens assures her that she would be supplied with the first copies of his son's work from then on for her reading pleasure. This promotion by Constantijn of his son's work speaks to the important role he played in cultivating relationships through book distribution. Presentation copies such as Princess Elisabeth's, gifted outside of normal market channels, allowed authors like Huygens to continue to work in a personalized intellectual economy, as they had when manuscripts and letters were shared in tightly formed networks. But sharing one's printed work had additional value beyond the cultivation of support.

In his Cutlerian lectures, Robert Hooke wrote about the need for scientists to publish in order to obtain credit for their ideas. "The securing of inventions to their first authors . . . tis hardly possible to do by any other means," and if an author prevaricates, says Hooke, he would find that there are spies who "think they do ingeniously to print them for their own."[25] Hooke, of course, was famous for getting himself deeply embroiled in conflicts over priority, but his words nevertheless ring true. To print in this period was to secure credit in a way that was widely recognized. After publishing his two mathematical works, Huygens shifted gears to focus on astronomy and mechanics. Beginning in 1655 he devoted himself to the technical aspects of astronomical observation, soliciting detailed instructions on lens-grinding from the best craftsmen in Europe. Within a year he had used his instrument to compile an impressive series of observations of Saturn.[26] In presenting his conjectures regarding Saturn's shape, Huygens employed numerous styles of publication: pamphlets, books, and letters—both public and private. His careful selection of publishing mode, coupled with strategic distribution, allowed him to target readers who were in a position to support his claims and confirm his place among respected astronomers. Both his 1656 *On Saturn's Moon* and the 1659 *System of Saturn* were sent to a select group of readers before they were available publicly; we therefore have a very good sense of the initial readership that Huygens considered appropriate for the works. But equally important, these astronomical works show us how an author could capitalize on personal distribution networks, and exchanges outside of general publishing corridors, to methodically and advantageously control the rollout of new ideas.

De Saturni Luna Observatio Nova was, as its title suggests, the announcement that Huygens had discovered a moon orbiting Saturn, what we today know as Titan. Like other astronomers he was inspired by Galileo's discovery of the moons of Jupiter and interested in Saturn's unique appearance, with its ear-like elongations visible through the telescope. Huygens speculated about the planet's shape with some correspondents but hoped to reserve his theory for later publication. He wrote astronomer Ismael Boulliau with a clear plea, "I ask you not to communicate to anyone what you know of the Saturnian world, nor to show the figure that I've drawn for you, until I've published the entire system."[27] In 1656, however, he had enough data to publish on Saturn's moon, which he had observed over the course of many nights. He enlisted the help of Adriaan Vlacq, a printer and bookseller from The Hague, who agreed to publish the work. Huygens received at least thirty-two copies of *De Saturni Luna* for his own distribution, and he restricted the audience to those who shared an interest in the topic, principally astronomers in Rome, Danzig, and Copenhagen, as well as members of the Parisian salon society, where he was hoping to foster relationships.

Distribution of the work can be traced through correspondence, and the level of control Huygens exercised in circulating the pamphlet calls into question exactly what constitutes a "publication." In correspondence he consistently used two distinct terms with respect to his works: *imprimée* and *publié*. He applied the former to *De Saturni Luna*. It was something he had printed but it was not, to his mind, published. Harold Love studied the liminal status of works like these, purposefully situated between the public and private.[28] He argues that the community of recipients defines how we should think about the work. In the case of *De Saturni Luna* we have a private publication for which the audience was entirely selected by Huygens, much like Tycho did with his astronomical works. On March 8, 1656, Huygens recorded distribution of the first copy of *De Saturni Luna*. It went to the Polish astronomer Johannes Hevelius via Huygens' brother Philip, who delivered the work while on a diplomatic trip.[29] Hevelius responded in June with thanks and a promise to reciprocate with his own contributions.[30] Thirty-one additional copies were sent, including one to the Parisian *savant* Jean Chapelain, along with a lengthy note from Huygens that refers to Chapelain's involvement in the publication process:

> When you see the pamphlet that I take the liberty of sending you, you will perhaps recall that it is by your counsel that it appears today. You believed it was important that I announce to those interested parties my new discovery concerning the planet which accompanies Saturn, and it is because of the zeal you have for the advancement of the sciences that they will be indebted to you so much: my aim was to delay the publication until that work was more perfect,

and to be content, in the meantime, to communicate it to certain of my friends who enjoy such things.[31]

In sending the work to Chapelain, Huygens understood that he was actually sending it to a much larger audience. On April 8, 1656, Chapelain wrote Huygens praising the pamphlet, assuring him that it would enhance his reputation as a serious astronomer. "I have viewed this publication with a particular joy; I have read it in our Academy assembly at the home of Msr. le Chancelier, and in another [meeting] of men of letters at Msr. Menage's, and at both the novelty of the idea has been pleasing and its beauty delighted those who attended. I hope that my attention has contributed something to the aggrandizement of your reputation."[32]

Chapelain confirms that copies were also passed on to Monsieur de Montmor, as well as to another, unnamed mathematician. He also vows to help in securing Huygens' honor in this area of study. Just as entering the network of Jesuit mathematicians had been a critical move in establishing Huygens' reputation in mathematics, earning a name as an astronomer among Paris' learned community—itself a tightly bound network—was the important first step toward recognition in the field. He would need his Parisian friends a few years down the road, when disputes arose over his findings. In the short term, however, Huygens began a lengthy series of conversations with colleagues about other Saturnian discoveries, all carefully orchestrated to allow certain individuals to comment on the findings while binding them to secrecy. "Of the system, no one knows anything except Mr. Boulliau, recently here [in the Low Countries], and perhaps it would be better if it was not divulged until one sees all the evidence of the work which I hope to provide soon," he writes Chapelain. And discretion among his correspondents prevailed when Huygens published his theory of Saturn's ring in 1659's *System of Saturn*. A small network of colleagues suddenly became an international audience.

Huygens circulated at least fifty-three copies of his 1659 book. While some recipients received a single book, others served as nodal points for distributing multiple copies. Ten copies, for example, went to Parisian astronomer Ismael Boulliau with instructions to deliver three to Jean Chapelain, four to mathematician Pierre de Carcavy, and one each to mathematicians Gilles Personne de Roberval and Claude Mylon. For individuals who received multiple copies, Huygens stipulated who should get them. At least six copies also went to the Low Countries, three to astronomer Johannes Hevelius in Danzig, and four to

Christiaan Huygens's list of recipients for his 1658 and 1659 treatises, *Codices Hugeniani. Courtesy of the University of Leiden.*

astronomers in London. The book also went out to a circle of Italian astronomers, lens grinders, and instrument makers—particularly those connected with the Florentine Cimento Academy—whose observational abilities earned them a strong reputation on the Continent.[33]

With patronage support from Leopold de' Medici, the astronomers affiliated with the academy were among the most important readers of Huygens' book; their positive response to his hypothesis would carry weight, and yet it proved the most difficult for Huygens to obtain. In part this was because an Italian astronomer, Eustachio Divini, had developed a theory of Saturn's appearance that challenged Huygens' ring theory. From Paris, Jean Chapelain tried to assuage Huygens' concern about the debate: "As for the Italians, who are opposing your beautiful work, and in truth with much baseness, I am not worried about it. And regarding the extract of the letter from Rome that you sent me, I don't see anything there against your position, but only that they want to try to find a fault with it, that they make the telescopes by this particular design and that they promise wonders from it."[34] In the end, after a back-and-forth of published responses between Huygens and Divini, Prince Leopold de' Medici declared Huygens the winner: "I read the work eagerly and attentively and I have shown it to many learned individuals, all of whom are pleased, seeing in it the solid characteristics of science and of your own excellence."[35]

In the case of Huygens' Saturnian publications, it is clear that targeted circulation allowed him to secure priority, not only by having his work published and sent directly to those whom he wanted as readers but by developing a cadre of supporters for his theory. When his ring theory—published only as an anagram in 1656—became public in the 1659 book, Huygens had the reputation and the audience he needed to engage forcefully with the Italian astronomers. His theory was right, but it would not have received the kind of hearing it did in the court of Leopold's scientific academy had it not been so well supported across the Continent.

In 1636 René Descartes was preparing to publish his *Discourse on Method* and three essays, *Meteorology*, *Optics*, and *Geometry*. His search for a publisher led him to Leiden, where—after considering the Elsevier firm—he chose Jan Maire to print his work. He writes Mersenne, "I would like to have the whole thing printed in handsome fount on handsome paper, and I would like the publishers to give me at least two hundred copies because I want to distribute them to a number of people."[36] A contract for three thousand copies was drawn up, a highly ambitious number for the time and one that gestures at

Descartes' confidence that the work would be of broad interest. Though he acknowledges that the *Geometry* "demands readers who are not only skilled in the whole of geometry and algebra . . . but also industrious, intelligent and attentive," he also believed that as a whole the work could be understood even by to those who had not studied the topics previously.[37] "I wished [it] to be intelligible in part even to women while providing matter for thought for the finest minds."

Descartes circulated his copies of the *Discourse* to numerous readers, including Jesuit mathematicians (many of whom were professors), physicians, theologians, natural philosophers, and notable political figures. With most copies he included a letter soliciting substantive comments on his ideas. "Make them as unfavorable to me as you can," he writes. "That will be the greatest pleasure you can give me. I am not in the habit of crying when people are treating my wounds, and those who are kind enough to instruct and inform me will always find me very docile."[38] We should, of course, read Descartes' willingness to hear criticism with some skepticism. Docility toward those who disagreed with him was seldom in evidence, but he nevertheless hoped that by selecting his audience he could encourage positive responses. Certain recipients also had the political cachet to open more difficult doors for Descartes' work. Cardinal di Bagno and Cardinal Barberini, who both received copies, were targeted not to elicit a substantive response but to smooth a pathway for the work's sale in Rome, despite the scarcely veiled Copernican ideas it posited.[39]

Four years later Descartes was again developing a treatise, this one philosophical. In an attempt to ensure his work was well received, Descartes decided to seek out intermediary readers—what we might call today early reviewers—who could provide feedback on his ideas. But as he completed the work, titled the *Meditations*, he delayed its printing because, he writes Mersenne, "I do not want them to fall into the hands of pseudo-theologians—or, now, into the hands of the Jesuits whom I foresee I shall have to fight—before I have had them seen and approved by various doctors, and if I can, by the Sorbonne as a whole . . . I did not want to have them printed until I was about to depart [for France] for fear that the publisher would steal some copies to sell without my knowledge, as often happens."[40] This sentiment reflects both Descartes' concern over publication and his creative effort to secure a positive response to his work. Two months later Mersenne received another letter, this one clarifying Descartes' plan for publication. I quote it at length because it reveals at once Descartes' thinking about the public, about printers, and about the importance of a favorable reception:

> I intended to have printed only twenty or thirty copies of my little treatise on metaphysics, so as to send them to the same number of theologians for their opinion of it. But I do not see that I can carry out that plan without the book's being seen by almost everyone who has any curiosity to see it; either they will borrow it from one of those to whom I send it, or they will get it from the publisher, who will certainly print more copies than I want. So it seems that perhaps I will do better to have a public printing of it from the start. I have no fear that it contains anything which could displease the theologians; but I would have liked to have the approbation of a number of people so as to prevent the cavils of ignorant contradiction-mongers. The less such people understand it, and the less they expect it to be understood by the general public, the more eloquent they will be unless they are restrained by the authority of a number of learned people.

Once the *Meditations* were printed, Descartes sent advance copies to the theologian Johannes Caterus, to the philosophers Pierre Gassendi and Thomas Hobbes, to logician Antoine Arnauld, and to faculty at the Sorbonne.[41] The recipients were expressly invited to respond to Descartes' arguments with their feedback. Through such a carefully managed release of his work, he found a window into the work's reception. In November 1640 he wrote to Guillame Gibieuf, a faculty member of the Sorbonne:

> One problem none the less remains, which is that I cannot ensure that those of every level of intelligence will be capable of understanding the proofs, or even that they will take the trouble to read them attentively, *unless they receive a recommendation from people other than myself.* Now I know of no people on earth who can accomplish more in this regard than the gentlemen of the Sorbonne, or anyone from whom I can expect a more sincere appraisal, and so I have decided to seek their special protection . . . I rely on your advice in telling Father Mersenne how he should conduct this business, and on your kind help in securing favorable judges for me.[42]

In this case Descartes' concern was obviously to avoid a religious backlash; by testing his ideas in limited release, he hoped to avoid the ire of the church. The examples of Giordano Bruno and Galileo, both of whom had been targeted by the church (and Bruno executed), were at the forefront of Descartes' mind. Nevertheless, his direct distribution of the *Meditations*—printed but not yet published—gave him access to scholarly readers' responses in advance of a wider circulation. It also afforded him the opportunity to publish his own replies to those scholars: "If my readers find difficulty at any point they will find it clarified in my replies."[43] Six individual meditations, the chosen readers' objections to them, and Descartes' replies to each were published together in 1641, coming off presses in both Paris and Amsterdam.

Descartes' attempts to secure a positive reading are mirrored in Christiaan Huygens' distribution efforts. In the introduction to his last published work, *The Cosmotheros*, Huygens quotes the Greek philosopher Archytas, "That tho a Man were admitted into Heaven to view the wonderful Fabrick of the World . . . yet what would otherwise be Rapture and Extasie, would be but a melancholy Amazement if he had not a Friend to communicate it to."[44] He goes on to articulate what it meant to share his scientific output:

> I could wish indeed that all the World might not be my Judges, but that I might chuse my Readers, Men like you, not ignorant in Astronomy and true Philosophy; for with such I might promise my self a favourable hearing, and not need to make an Apology for daring to vent any thing new to the World. But because I am aware what other hands it's likely to fall into, and what a dreadful Sentence I may expect from those whose Ignorance or Zeal is too great, it may be worth the while to guard my self beforehand against the Assaults of those sort of People.[45]

His reference to "those sort of people" rings with condescension. We might, then, dismiss this entire passage as typical rhetoric, with its inherent flattery: "You, holding this book, are my ideal reader," he suggests. But Huygens and his contemporaries actively—and repeatedly—pursued specific readers for their works. We should therefore read their statements about audience as expressions of a best-case scenario, if not of a reality. Huygens hoped to avoid the readers who might come by his work in a bookseller's stall and read it without the proper foundation. His reluctance to suffer the "dreadful sentence" of the unprepared reader echoes sentiments expressed before him by René Descartes, and his lists of recipients for presentation copies can be thought of as templates for avoiding such a fate.

There were, of course, gradations of readers in this period, as Huygens' lists attest. Elite scientists such as Newton viewed themselves as the creators of knowledge; they produced the majority of the ideas and theories. A step down from them we see amateur scientists, gentlemen with the means, aptitude, and interest to understand scientific discoveries and debates, but often without the skills to lead in any particular field. John Locke is a good example, a keen intellect and supporter of Newton but hardly a mathematician. The French *philosophe* John Desaguliers claims, "The great Mr. Locke was the first who became a Newtonian Philosopher without the Help of Geometry." Such a characterization hardly sounds charitable, but Desaguliers was capturing a kind of reader who proved important in early modern science. He continued, "Having asked Mr. Huygens, whether all the mathematical Propositions in Sir Isaac's Principia were true, and being told he might depend on their Certainty; he [Locke] took them for granted, and carefully examined the Reasonings and Corollaries drawn from them, became Master of all the Physicks, and was fully convinc'd of the great Discoveries contain'd in that Book."[46] Finally, beneath the strata of elite and amateur readers there was an interested audience of educated readers, a group fascinated by the sciences but with limited ability to completely grasp the technical details. This lay audience, while enthusiastic, could seldom summon the resources or the skills to contribute meaningfully to a given scientific field, preferring instead to enjoy ideas at the level of entertainments.[47]

If we look at Huygens' publications on the mechanics of the pendulum clock, we see how he viewed these multiple audiences and the way he circulated his works to cultivate favor among each of them. His first publication on the pendulum clock was the *Horologium*, printed in The Hague in 1658 and distributed privately to at least fifty-nine individuals. Huygens' aim with the work was to make his new technology widely known. He sought priority not only for himself but for the Low Countries as well, dedicating the book to the "the Most Illustrious and Most Powerful Members of the State of Holland and of the Western Region," the governing body of the Low Countries. His dedication continues, "I felt myself strongly impelled to ensure to our country the credit for this and for any future discoveries, and so I have followed the way which alone seems proper to this end—*to make known the whole idea and construction of the new mechanism*, which I, the inventor himself, have undertaken to describe in a few words and to produce to the public in a reasonably brief volume."[48] Under the heading "Exemplaria Horologij, primum editionis," Huygens lists in his workbook fifty-nine recipients of the publication. The names are loosely grouped according to nationality. Topping the list is the States General, the governing body of the Low Countries that had approved a patent for Huygens' clock on June 16, 1657.[49] As dedicatees of the work, they were also given a pendulum clock by Huygens, which was hung in the chamber where the representatives met. Following the States General is Johan de Witt, councilor pensionary of Holland and former student of van Schooten at Leiden, where he met Huygens. De Witt supported Huygens' efforts to obtain patent rights for various clock models.[50] A number of recipients from the Low Countries are below de Witt on the list, followed by French and English recipients. Huygens' list was a clear example of circulating works to secure priority, but what followed with the second edition of this book is a stark contrast.

In 1666 Huygens charted a new course in France, where he became a leading member of the newly founded Académie Royale des Sciences. The move to Paris was welcomed not only by Huygens but by his many correspondents there, especially Jean Chapelain, who had remained a close ally since his publication of the 1656 work on Saturn. It was Chapelain who continually pushed Huygens to get his Saturnian ring theory into print, and his advice was similar with regard to the horological work. From the time the 1658 *Horologium* appeared, Huygens had characterized it as a first edition, and Chapelain repeatedly—one might say doggedly—inquired about the progress on a revised, second edition, which finally appeared in 1673. As was his custom, Huygens sent copies of the *Horologium Oscillatorium* to individuals throughout Europe. In May, Huygens wrote to the secretary of the Royal Society, Henry Oldenburg, "It is now already some time ago that I sent you a dozen copies of my book on the clock . . . and I ask you kindly to take care of the distribution of all the books according to the inscriptions which I have put in them."[51] Oldenburg's letter confirming receipt of the copies arrived in Paris two days later. He told Huygens he had distributed them according to the inscriptions, noting that, "when these gentlemen whom you have entertained with your book have read and pondered it, and thought good to share their opinions of it with me, you shall not fail, Sir, to be informed."[52] And even though Huygens' father had ceased to be a presence on his son's title pages, he remained keenly interested in how people responded to Christiaan's work. In his own letter to Oldenburg Constantijn writes, "Mij french sons new oscillatorium I believe by this time hath been seen amongst you. For as much I may claim to be the grandfather of that childs-child, I doe long to heare the R. Society good opinion of it, and specialy the judgment of your most learned and worthie Presid. the Lord Brouncker and the Illustrious Mr. Boyle whose wonderfull capacity and universal knowledge in *omni scibili* I doe still admire and little less then adore."[53]

Among the recipients at the Royal Society were high-caliber mathematicians like Isaac Newton, John Wallis, Lord Brouncker, and Christopher Wren; natural philosophers such as Robert Boyle; instrument specialists like Alexander Bruce and Robert Hooke; and lay readers, as exemplified by Oldenburg. Hooke was a particularly important recipient because, as Huygens well knew, he claimed to have produced a conical pendulum prior to Huygens. In his diary entry of May 30, 1673, just after receiving his copy of the *Horologium Oscillatorium*, Hooke writes, "Saw that he [Huygens] had taken my Invention of circular pendulum and for falling bodys."[54] He reached out to Constantijn Huygens and Henry Oldenburg to register his complaint, but it was clear from Huygens' lengthy discourse on the mathematics of the pendulum that he had secured credit for this particular achievement. Overall, Huygens' 1673 treatise would be read differently by each of the subgroups he identified, but no matter what their background, each would find part of the work accessible to them, and Huygens gained something—status, support, praise—from each reading.

A gap of eleven years, and a great deal of political turmoil, separated Huygens' horological publications from his 1684 *Astroscopia Compendiaria*, a book that introduced Huygens' newly invented tubeless telescope. A central problem in astronomy in the seventeenth century was the presence of spherical and chromatic aberration, distortions that resulted from light rays not being brought to a unique focus. Improvement in the telescopic image required a long focal length, which in turn meant a very long telescope. While most were between 30 and 50 feet, Johannes Hevelius designed a telescope that reached 150 feet in length. But managing a telescope of that length was cumbersome. It was difficult for astronomers to get the correct alignment between the focal and objective lenses.[55] To make these telescopes more practical, one needed to eliminate the tube, so that the eyepiece and objective lens of the instrument were not encased. That way they could be mounted as far apart as one desired and moved independently to get the best possible alignment. Huygens, living in the Hague since the revocation of the Edict of Nantes, did just this, and he hoped his new invention would help restore his support in France.

In *Astroscopia Compendiaria*, which was composed of just twelve pages and one plate, Huygens describes his tubeless telescope with moveable objective lens. But this publication, sometimes referred to as a pamphlet or brochure, took a unique turn as it came off the press of Arnold Leers, a printer in The

Hague. Huygens explains to his readers in a preface (composed *after* the pamphlet itself), that the original *Astroscopia* had been "printed but not yet published . . . when, as it often happens, another idea presented us the means to make our invention even better and more useful."[56] The improvement required additional text and an illustration of the telescope's new design. The textual piece came in the form of a direct address, "To the Reader," and in it Huygens explained the new mechanism for holding the objective lens.[57] The added engraving was done by a Mr. Tarpentier, who had—Huygens noted—capably done the original plate for the work. The good news was that copies of the *Astroscopia* that were meant for the general readership were not yet available from booksellers. Huygens therefore anticipated that his addendum could be attached to the original printed pamphlet in reverse chronological order. Composing it as a letter to the reader effectively guaranteed this would happen.[58] Only after the preface did readers encounter the work he had written originally, which included a description of the telescope and a plate showing its setup. This assumes, of course, that they purchased the pamphlet from a bookseller. Those who were sent copies directly from Huygens had a slightly different reading experience.

By May 1684 Huygens had distributed copies of the *Astroscopia* to dozens of individuals. Thus, when he discovered an improved design, he had to send the printed addendum to the same group of recipients. From correspondence we know there was about a one-month gap in between the mailings. A complete list of the forty-one individuals who received presentation copies of the *Astroscopia* can be found among Huygens' papers.[59] Curiously, only three went to academicians in Paris—Giovanni Cassini, Claude Perrault, and Jean-Baptiste Duhamel—a surprisingly low number for a work on astronomical instruments, especially given that Huygens had lived and worked among so many astronomers there. Huygens instructed Perrault to share the book with several other members of the French academy until such a time that he could send out more copies. He was not discounting his former colleagues, but his primary concern was earning the favor of a powerful new figure in French politics, François Michel Le Tellier, Marquis de Louvois. Though nominally the secretary of state for war, Louvois wielded significant influence in many quarters of government, and after the death of Jean-Baptiste Colbert he was named prime minister to Louis XIV. Louvois' general apathy for the work of academicians— not to mention his anti-Protestant inclination—left men such as Huygens without the support they needed.[60] What interest Louvois did have in the sciences tended to focus on research in natural philosophy, which received over half of the academy's budgetary allocation.[61] Where Colbert had committed 39,650 livres a year to the Paris Observatory, Louvois allocated only 1,371 livres, a fact that did not bode well for Huygens. In a letter to Louvois, Huygens requests that the minister "be kind enough to send me your commands." There was no reply. His father, Constantijn, also attempted to reach out through a Parisian friend to determine Louvois' intentions vis-à-vis his son, but he, too, heard nothing. Finally, in May 1684, Huygens sent the minister a copy of the *Astroscopia* and repeats that he is "recommending myself to your attention and still awaiting the honor of your commands."[62] Huygens heard nothing. "The reason for this decision is unknown to me, but I must resign myself to it." His tenure at the French Academy had come to an end; his career, however, had not, as he continued to collaborate with colleagues on the continent and in England.

The largest group of *Astroscopia* recipients were members of the Royal Society. Huygens sent four copies to Daniel Colwall, then vice president of the society. This fundamental reduction in presentation copies to French academicians and Royal Society members is perplexing, on first pass, but it signals an important shift in the publication landscape. Huygens still sought to steer his work toward the scientific community, but by the latter decades of the seventeenth century he had a new and effective option in the form of academic journals. Contrary to what we see with his previous publications, the ideas presented in Huygens' *Astroscopia* were distributed principally through scholarly journals, some of which had been in print for more than a decade, while others were relatively new. Huygens was keenly aware that instead of sending copies to individuals, he could target journal editors who might give the work a favorable review. Whereas the distribution of single copies helped foster relationships with individual readers, a summative journal review could influence scores of readers at once, and they would likely be his ideal readers. It is interesting, then, that Huygens—who had long attempted to define his audience with dedication copies—ended up using the printed journal as a means of managing the wider readership that print culture offered him.

As historians have noted, the rise of the academic journal marked a profound change in the way scholars communicated. The French *Journal des Sçavans* appeared first in 1665, a twelve-page pamphlet that purported to "make known all that is new in the Republic of Letters."[63] Its editor, Denis de Sallo, promised readers the latest discoveries in the sciences, synopsized content of recently published scientific books, and obituaries of celebrated men, all of which he deemed of interest to the Republic of Letters. He imagined the journal as an international forum through which scholars could collaborate and compete. In the same year, the *Philosophical Transactions* was published by Henry Oldenburg on behalf of the Royal Society. In his periodical reports of experiments and sightings, Oldenburg included firsthand accounts of natural phenomena culled from his vast correspondence and notices of recently published books.[64] These journals were followed by others, including the *Acta Eruditorum*, published in Germany starting in 1682, and the French *Nouvelles de la République des Lettres*, published in Amsterdam in 1684. Traditionally these journals have been viewed as democratizing media, vehicles through which scientists could communicate their ideas to a broader audience in a more efficient way. They represented the fusion of Baconian technology and ideology, where the printing press would facilitate a transparent presentation of ideas, observations, and theories for a public who could then engage in a debate about them. Such a utopian vision, of course, was not fully manifest in practice. As with printed books, journals presented scientific authors with both opportunity and liability. There were several concerns, starting with the frequency of journal publication, which some believed would contribute to the growing sense of information overload.[65] Furthermore, there was concern that the excerpting and synopsizing of books would lead to laziness among readers and—arguably worse—would encourage dabblers. As we saw in previous chapters, authors took seriously the need to have well-informed and skilled readers. The journal threatened to undermine this by encouraging dilettantism. As Thomas Broman succinctly puts it, journals allowed "matters of fact" to be turned into "factoids" for the public.[66]

These risks, however, were exceeded by the advantages journals presented to scholars, the most obvious of which was the ability to connect to a relatively narrow and educated audience. Journals also provided a forum in which scientists could engage in sustained debate or develop a line of investigation over the course of several issues. Nearly a dozen reports of experimental blood transfusions, for example, were published in the *Philosophical Transactions* and the *Journal des Sçavans* between 1667 and 1668, with anatomists from London and Paris sharing observations and hypotheses, challenging each other's accomplishments, and attempting to take the lead in determining the effects of transfused blood in animals and humans. Articles that appeared in one journal would often be translated and reprinted in the other, as when the *Journal des Sçavans* printed a translation of the Royal Society's announcement of a successful transfusion.[67] A similar dynamic is seen with published reports of astronomical phenomena, which were observed at different geographic locations and then compared in the journals. Scientific periodicals fostered dialogue between scientists in a way published books did not, and they stirred competition among practitioners in specific fields. Finally, there was the practical benefit of book reviews, a new and necessary tool in the face of so many published works. Though the sheer number of reviews would eventually become as oppressive as the books themselves, at the outset they provided authors with a valuable device to help sift through new material. An astronomer in London could pore over the latest titles published not only in England but on the Continent as well, determining which were worthy of tracking down from local or foreign booksellers. Toward the beginning of the eighteenth century, the critics who authored these reviews came to be recognized mediators between the specialized work of scientists and the lay public that was keen on understanding the ideas but lacked expertise to read the work themselves.[68] On the whole, men like Christiaan Huygens recognized the advantages of these scholarly journals and embraced their use. Journals acted as a newer and nimbler version of the correspondence networks that had sustained scientific communication for so long.

On his 1684 list of recipients for presentation copies, Huygens included Jean-Paul de la Roque, then editor of the *Journal des Sçavans*. In a letter from June of that year Huygens asked de la Roque to publicize the *Astroscopia* in the French periodical: "I have sent an example to Mr. Cassini and another to Mr. Perrault the physician, who will have the goodness to loan it to you in order to take an extract, if you would take the trouble to do so. In order to avoid making this package too large I am sending you only the figure with a summary explanation. After the printing I found a considerably important addition to this invention, which I will send you as soon as possible after having it engraved."[69] On December 4, 1684, an article on Huygens' book appeared in the *Journal des Sçavans*. De la Roque, writing on behalf of the academy, notes that academicians had sought ways of improving the telescope over the past year, and that Huygens has accomplished this by developing an instrument with a long focal length but without a tube. While Huygens's full explanation is not recounted in the piece, the *Astroscopia* plates are included, which de la Roque says should be sufficient for understanding the invention, at least to those familiar with telescopes. It is also clear in the title of the review that while the book was published in The Hague, copies could be found in Paris. In addition to the academy's periodical, Huygens wanted his work discussed in *Nouvelles de la Republique des Lettres*, which Pierre Bayle had started earlier that year.[70] In style it resembled the *Journal de Sçavans*, but substantively it consisted solely of book reviews that Bayle authored himself, which explains why it was a monthly, rather than a weekly, publication. Bayle was committed to the idea that such a journal could help create a pan-European republic of letters, composed of the casual learned reader and the *érudit*. He worked doggedly to achieve that aim, authoring five hundred reviews in just three years.[71] A review of the *Astroscopia* appeared in the May 1684 issue.[72] Huygens' pamphlet was also discussed in the international journal *Acta Eruditorum*. Founded in Leipzig in 1682 and edited by Otto Mencke, it quickly became a respected Latin-language periodical. In structure and content the *Acta* was very similar to the *Journal des Sçavans* and *Philosophical Transactions*, often sharing articles with those journals. Between eight hundred and a thousand copies of each volume were printed and distributed to booksellers across the Continent. Having one's work reviewed there meant a significant level of circulation among Europe's learned community. The December 1684 issue of the *Acta* offered an extensive description of Huygens' telescope written by Christoph Pfautz, though as was common, the article was published anonymously.[73] Finally, the *Philosophical Transactions* of July 1684 ran a three-page review of Huygens' book.[74] In a series of seven steps, the reviewer—most likely the astronomer Detlev Clüver—details some of the challenges encountered when using a telescope and then describes how Huygens' telescope circumvented these problems. With around one thousand copies printed of each issue of the *Philosophical Transactions*, Huygens again saw his work reach an expanded readership. More importantly, his efforts to reach four separate journals to promote his book reflect the extent to which publication strategies were changing. Fewer individual copies needed to be circulated to the natural philosophical and academic communities, who would become aware of the work through reviews. What copies Huygens did personally circulate went instead toward a lay audience: diplomats, gentlemen, and friends.

Journals proved important again in 1690 when Huygens published what would be his penultimate work, *Treatise on Light and Discourse on the Cause of Gravity*. Huygens sent copies for review to the *Acta Eruditorum* and the *Journal des Sçavans*, and the work was reprinted in *Mémoires de mathematique et de physique de l'Académie Royale des Sciences*, a compendium of treatises from the

French Academy.[75] A copy also went to Pierre Bayle, despite the fact that he was no longer publishing his journal, along with a letter from Huygens: "You are, Sir, among the few men for whom this book was made, and whom I hope to have as readers. Because of this I am obliged to send you this example."[76] On February 7, 1690, Huygens wrote Fatio de Duillier, London mathematician and close colleague of Newton, to describe his *Traité de la lumière* and to apologize for how long it took to produce. He had hoped to send it months earlier: "I was wrong and I found myself extremely mistaken about the slowness of the printers, who had me waiting four months and counting. So here at last I send to my brother of Zulichem, Secretary to the King, seven copies of this book, which I ask him to give you along with this letter. They are for you, Sir, and for Msrs. Newton, Boyle, Hamden, Halley, Locke and Flamsteed, all of whom you know and most your good friends. I dare ask you to ensure their distribution to them."[77] Huygens explains that there were several parts of the book—particularly those treating gravitation—that he thought would be relevant to Newton: "I would be pleased to know his sentiments on my explanation of refraction and the phenomena of Iceland crystal. But I am not sure that he understands French. I wish above all to know how it seems to you, Sir, who are the most competent judge that I know of in these matters."[78] On the same day Huygens informed his brother Constantijn that eight copies of the *Traité* were being delivered by messenger, along with details on recipients.[79] A few weeks later Constantijn wrote to request additional copies for Lord Pembroke, then president of the Royal Society, as well as for Christopher Wren and John Wallis. Copies were dispatched immediately. Huygens had been elected to the Royal Society in 1663 and spent considerable time with society members during his 1689 trip to London. This was particularly important in light of his departure from the French Academy. Meetings of the Royal Society throughout 1690 saw vigorous debate over Huygens' theories of light and gravity, with Robert Hooke volleying criticisms while Edmund Halley mounted a defense of the works.[80]

Individual responses to Huygens' presentation copies reflect deep engagement with the works, especially the treatise on gravity. Fatio de Duillier received his copy on February 10, 1690, and immediately wrote to Huygens, "I have read through your Treatise on Light with singular pleasure, and I have already read the part on gravity several times."[81] He would later send five pages of comments to Huygens for consideration. As for Isaac Newton, who clearly had an interest in the work on gravitation, de Duillier reports, "I have found him so willing to correct his book [the *Principia Mathematica*] on those items which I mentioned that I could not admire any more his ability, particularly in those places where you attacked him. He has some difficulty understanding the French but he manages with a dictionary." On the Continent, copies went to readers in Leiden, Copenhagen, Marburg, Hamburg, and Paris. One recipient was Leibniz, whose copy arrived along with a letter apologizing for the delay, which Huygens again attributed to the slow pace of the printers. Huygens told Leibniz that the "large volume" he was sending would require "several hours of your time to be read," but "as a competent judge of these matters . . . I would ask only that when you have examined these brief treatises, that you let me know your thoughts and whether I have been too bold in advancing something which seems new to you and satisfies you."[82]

Leibniz was happy to receive the copy from his Dutch colleague, particularly since the book was not yet available from Hamburg booksellers. At the same time, in Paris, the academician and astronomer Philippe de la Hire expressed regret at the *Traité* not being readily available in France. "If our booksellers had been able to get your work on light and gravity, I would have already seen it, but one must have a little patience."[83] Two months after this exchange, Huygens' treatise had still not arrived in Paris. De la Hire continued to write Huygens between May and August to inquire about the book, saying academicians were eager for the work.[84] Meanwhile, Huygens sent his French colleagues copies of the *Astroscopia Compendiaria*, which he said would give academicians something to read in the interim. While he was waiting, de la Hire expressed his concern to Huygens about the arrangement to distribute books to certain individuals:

> I have not been able to prevent telling some of our friends of the new things that you have sent me about the printing of your book, such that the thing is divulged. I found it somewhat uncomfortable that because you send the books only to certain people, many others, themselves members of the *Académie*, are chagrinned because you have omitted them. What surprises me most is to see that you don't send one to Chapelle, who always has your interest in all affairs. It is for this reason, Sir, that I find myself obliged to not conduct this business with each person, [but] to put [the books] in the hands of one of our booksellers and to inform those to whom you addressed them to go retrieve them from that place directly, because one would have from this trouble the knowledge that it was not me who would make this distribution up on my own.[85]

This is the first time any of Huygens' chosen recipients expressed reluctance about his means of distribution. Individuals who had previously served as nodal points for distribution, such as Jean Chapelain, Ismael Boulliau, or Fatio de Dullier, carried out the distribution of books as Huygens requested with no apparent reservations. De la Hire's reticence may have stemmed from his awareness that knowledge was being shared in very different ways at the end of the century, and that Huygens' selective distribution strategy was passé. Nearing the eighteenth century, scientific communications were typically channeled through journals and books accessible to both scholars and the wider public. Thus, Huygens' attempts to target recipients with gifted copies was reminiscent of an earlier era, when patronage was leveraged with gifts and dedications. But de la Hire's discomfort more likely reflects his misgivings about the names on—or absent from—Huygens' list, specifically Henri de la Chapelle. De la Chapelle was a close confidant of Louvois, "his stooge in the Academy," according to Cornelis Andriesse, and adamantly opposed to Huygens' reinstatement.[86] His absence on Huygens' list of *Astroscopia* recipients was conspicuous; consequently, de la Hire felt compelled to seek an alternate means of distribution, hoping the task could be handed over to a bookseller. Huygens, dismissive of de la Hire's concerns, replied directly: "The names of those to whom the examples go are marked in such a way that it will be nothing implicating you during the distribution."[87] Ready to take full responsibility for his selection, he pushed de la Hire to go ahead with the distribution as requested. When the books finally arrived, de la Hire informed Huygens that he had acted as instructed. His comments are revealing about the work's reception in Paris:

> I am extremely obliged to you and believe that you will receive the compliments of those to whom you have sent copies. By the time the packet arrived I had read a copy of your work that was in the Royal Library and which had come by post about six months ago . . . [Y]our treatise seems to me altogether a masterpiece of physics and mathematics, and I know to allow myself to admire the turn that you have taken to explain such extraordinary phenomena and to explain everything as you have. I have had occasion in the public lessons that I give at the Royal College to explain your system and I have argued it as well as is possible. It is a new thing in these quarters because except for the examples you have sent and the one in the Royal Library I do not think that there are any to be had in Paris.

Professors like de la Hire frequently offered special lectures in mathematics that students could not get elsewhere, such as those on Huygens' *Treatise on Light*. Huygens was not directly targeting students, but to reach them through a capable intermediately was no doubt advantageous. Despite some initial reservations, de la Hire proved a receptive reader and an important advocate within the Parisian community.

Huygens' lists of his chosen readers are interesting for what they tell us about this particular scientist and his perceptions of the ideal audience. But they also provide critical insight into broader issues of early modern authorship. Huygens, like most scientists, understood that publishing a book was the primary way to present new ideas to the world in the seventeenth century.[88] Journals and pamphlets had a role to play, but books were imbued with an authority that endured, which those other genres were only beginning to obtain.[89] If Huygens' books were released to the public only through traditional outlets—the stalls of booksellers in London or the publishers' tables at the Frankfurt Book Fair—he would have failed to gain the attention, priority, or support he sought. His lists of recipients afford us a glimpse of the diverse audience he wanted, and, coupled with correspondence, we get a remarkably clear picture of how Huygens benefitted from targeting different recipients. By circumventing the normal distribution channels of printers, Huygens, Descartes, and other early modern scientists found a means of placing their works, their ideas, in the hands of the most skilled and advantageous of readers.

Chapter Four

"A True and Ingenious Discovery"

New Print Technologies and the Sciences

> Dr. Pell told me that somewhere in Porta was a way delivered of stamping
> letters & figures in Sallow and afterwards plaining them almost out then
> steeping or boyling them in water whereby the letters would swell outwards
> and be fitt for printing or ye like or be embossed with heads or the like.
> —ROBERT HOOKE (JANUARY 27, 1673)

W hen we think about early modern scientists engaged in the craft of
printing, the iconic figure of Tycho Brahe quickly springs to mind.
Brahe was a Dane of wealth and standing thanks to the patronage of King
Frederick II who, in 1576, granted the astronomer land on the island of Hven
and resources to build a home and an observatory there. In addition to craft-
ing some of the largest and most precise astronomical instruments of the day,
Brahe ordered the installation of two printing presses in his home. On November 27, 1584, he noted the completion of the first press. The inaugural
print job consisted of poems Brahe dedicated to friends, but by the end of 1585
he was printing a meteorological calendar and astronomical observations.[1]
That same year he requested and received an indefinite privilege to print from
Rudolph II, a type of ongoing royal permission assigned on behalf of the Holy
Roman Empire.[2] And after years of trying to maintain a supply of paper for his
press, Brahe had a paper mill built, a project that involved tremendous engineering effort, including building dams to generate the water power required.
He proudly wrote to a friend about the mill when it was finally completed:
"A high, wide embankment regulates the water supply, which suffices in summer as well as in winter. The wheel, which is approximately 7 meters in di-
ameter, is powered by the least possible amount of water, and besides the
manufacture of paper, is the source of power for two industries. A number of
fishponds are also laid out, so that these may also supply water for the mill
when it is needed. And only a few years ago, this was all just dry land."[3]

The mill was operational by 1591, producing paper for printing, and—if Tycho's description is understood properly—to prepare parchment as well. This parchment "mill" was powered by the same overshot wheel that drove the paper mill, and Tycho likely used it to prepare skins for bindings.[4] Coupled with Brahe's bindery, the entire publishing operation was brought to Hven.

Both the printing press and the paper mill were necessities to Brahe. Printers in Copenhagen were often too busy with other projects, and it was far more efficient for him to print his own works than to shuttle galley proofs back and forth to his residence at Uraniborg for correction. Brahe also held an abiding belief that, depending on the field, his work should not be widely shared. Astronomical observations, for example, could be available to anyone interested, but chemical treatises demanded a measure of protection:

> I shall be willing to discuss these [chemical] questions frankly with princes and noblemen, and other distinguished and learned people, who are interested in this subject and know something about it, and I shall occasionally give them information, as long as I feel sure of their good intentions and that they will keep it secret. For it serves no useful purpose, and is unreasonable, to make such things generally known. For although many people pretend to understand them, it is not given to everybody to treat these mysteries properly according to the demands of nature, and in an honest and beneficial way.[5]

Such a sentiment might suggest that Brahe preferred small print runs in order to tightly control readership, but the evidence complicates this assumption. Brahe printed fewer than one hundred copies of his *Astronomiae Instauratae Mechanica* of 1598, the book in which he described his many astronomical instruments, but such limited runs were not the norm. Fifteen hundred copies of *De Recentioribus Phaenomenis* were printed in 1588, and correspondence indicates that Brahe aimed for a similar number with his other books, whether they were printed at Uraniborg or not.[6] And while his books did not follow typical pathways from printing house to bookseller, Tycho still participated in the international book trade, sending copies of his works to the book fair in Frankfurt, and making them available to certain booksellers in markets where he believed readers would be interested. Ultimately, he wanted to see his works distributed, but, as we saw in the previous chapter, he hoped to guide that distribution in specific ways.[7] He sent copies to princes, aristocrats, and scholars across Europe, precisely the people he wanted to see his work, be it for intellectual recognition or patronage. For Brahe, an in-house press was one of the most effective ways to target his desired audience and control the distribution of his work.

The "Lord of Uraniborg," with money and means, is often seen as exceptional in his efforts, an unusual figure whose typographic achievements hardly represent broader trends. But scholars have recently questioned such a position. Rather than casting him as an idiosyncratic figure of the Scientific Revolution who stood outside standard printing conventions, he is better viewed as an astronomer whose publication practices reflect broader trends of the period, only in a greatly enhanced manner. Other scientific practitioners attempted to wrest control of printing from the commercial houses just as Brahe did. A few had the resources to install commercial-grade presses and to hire the staff required to operate them. Most worked more modestly, developing original printing or reproduction techniques that could produce as few as two copies of a manuscript, a dozen or more prints, or a series of engraved images.

Assuming the job of printing by literally taking it back from print shops seems to have been a minor trend, a mere hobbyist's effort along the lines of lens grinding or clock making. To be sure, most printed works came from the presses of established printing houses; there is no arguing that individual printing operations approached commercial levels. But we should not assume that the efforts to become involved with the world of print—which Brahe and his contemporaries certainly did—were entirely marginal, or even trivial. Scholars have repeatedly demonstrated that the divide between the intellectual and the craftsman in the early modern era was decidedly blurry. Knowledge held by artisans and craftsmen informed scientific thinking in myriad ways.[8] Instrument makers in early modern London, for example, cannot be singularly classified as scholars or craftsmen; rather, the relationship between the fabricator, the individual who offered instruction in the instrument's use, and the consumer was a dynamic one.[9] Likewise, gentlemen delved into the artisan's world, tinkering with watches or building scientific instruments to satisfy their curiosity and—at times—to provide some amusement. The famous diarist Samuel Pepys, for example, recorded a visit with the president of the Royal Society—Lord Brouncker—that illustrates this kind of activity: "I to my Lord Brouncker's and there spent the evening by my desire in seeing his lordship open to pieces and make up again his watch, thereby being taught what I never knew before; and it is a thing very well worth my having seen, and am mightily pleased and satisfied with it."[10] Reports like this are common; the thrill of using a planetarium, electrical machine, or a chronometer was widely shared. Many in the scientific community engaged with instrumentation quite seriously, applying both their talents and their means to develop new mathematical, experimental, or astronomical instruments to advance specific scientific agendas. René Descartes' studies of optics led him to design a special lathe to grind hyperbolic lenses that could be used in improved telescopes.[11] Robert Boyle's air pump was integral to his program of natural philosophy. He detailed its design and function in his publications so that others could replicate his experiments. And Christiaan Huygens delved into telescope design with great success, grinding lenses with his brother to build the instrument that allowed him to observe Saturn's moons and ring. His treatise on grinding glass reflected his intimate knowledge of the craft and presented the reader with details about lens shapes that he had gained through extensive experience. It was not a mathematical treatment of optics but rather a practical guide to the craft of grinding glass.[12] And finally, there was Robert Hooke, who devoted a career to developing new instruments for various lines of experimental research, both as a private individual and on behalf of the Royal Society, where he was curator of experiments. Hooke spent as much time with his colleagues at the Royal Society as he did among London's various instrument makers, men whose artisanal work placed them in a liminal position between the learned and craft communities. Jim Bennett, who has devoted a lifetime to studying instrumentation in this period, notes, "There is no sharp division between makers and mathematicians—some of the most able and prestigious among them, such as Johannes Regiomontanus, or Gerard Mercator . . . were happy to make instruments, to present themselves as makers, or to preside over workshops and printing presses."[13] Throughout London the shops of instrument makers were frequented by amateurs and established scholars who sought to purchase apparatuses that would assist them in their scientific work.

It is in this context of sustained professional interest in instrumentation that we should consider the development of small-scale printing techniques, the kind of operations that sat outside of traditional printing houses. Scientists from a variety of fields were thinking about new ways to print and how they could do it themselves. In reflecting on these efforts, Adrian Johns notes that "innovative appropriations of the very mechanisms of print helped make for important achievements in the sciences themselves."[14] Such luminaries as Johannes Hevelius, John Pell, Christopher Wren, Christiaan Huygens, William Petty, and Robert Hooke all developed new methods—new "technologies" as they called them—that would allow for small-batch and custom printing. They also experimented with etching and engraving techniques in an effort to improve the quality of printed images related to their work. Their sustained efforts amount to something more than an artisanal hobby, as historians of science and print have recently shown.[15] But beyond curiosity about the technical process of printing, why do we see so many scientists developing new ways to print? The answers are varied, but many, I believe, relate to the fundamental issues of control that echo throughout this book. A scientific author who printed their own material enjoyed, first and foremost, emancipation from the larger systems of communication technology, from the printing house and all the complications and compromises that implied. Self-printing also allowed authors to control who had access to their work, especially by reproducing it in small numbers instead of larger, commercial print runs. In this case, self-printing was equivalent to self-publishing, allowing an author to circumvent the booksellers' stalls and distribute their work in targeted fashion. But it also meant, in the case of both text and images, that authors could shape the way their work appeared on the page, ensuring that words and images, especially engravings, looked precisely the way they wanted them to and were integrated in a fashion they judged most effective. Matthijs van Otegem, for example, has considered the ways that typographical layout related to cognition in early modern scientific works. Images and text were placed on the page in complementary or counterintuitive ways; detailed text was supported by either stylized or didactic illustrations, depending on how the author wanted the reader to comprehend the ideas.[16] This was, van Otegem argues, closely related to new theories of cognition under discussion in the seventeenth century. When Martin Lister was preparing a natural history treatise and needed images done of some fossils he had collected, he enlisted the York engraver William Lodge. Over the course of several exchanges, Lodge explained to Lister that he rendered the fossils into images in two ways, "one as it apeares with most advantage plane wise" and the other "as it is erected perspective wise."[17] Lister was pleased with the results and included the images in his paper on fossils, which he sent to the Royal Society in 1673 and subsequently published in the *Philosophical Transactions*. Addressing the illustrations in his work, Lister told Oldenburg, "Words are but ye arbitrary symboles of things, & perhaps I have not used ym to ye best advantage. Good Design (& such is yt I send you, done by yt ingenious young gentleman & excellent artist, my very good friend Mr. William Lodge), or ye things ymselves, wch I have all by me, would make these particulars much more intelligible and plain to you." For Lister, the objects themselves had the greatest didactic value; good engravings were the next best thing and words, clearly, a poor substitute.

In addition to being more effective at transmitting information, images engraved by the scientist who made the original observations tended to accrue more authority. The categories of observer, draftsman, and engraver would effectively collapse, allowing the reader to trust the images more than they otherwise might.[18] Different disciplines, of course, demanded different qualities from their images, leading to debates about whether a plant, a comet, or a living body should be depicted as it exists in nature or whether, as some authors argued, the illustrations should be an "average" or essential depiction, which might be more useful to the reader.[19] Ultimately, scientists engaged in the technology and processes of print to escort ideas from the point of creation—observation, experimentation, calculation—onto the page and out into the world.

Like Tycho Brahe before him, Johannes Hevelius was an astronomer of significant wealth, having inherited from his father a successful brewery.[20] At his house in Gdańsk, Hevelius built a library, museum, instrument workshop, and a world-class observatory. In addition, he installed and operated his own printing press, even learning the engraving process in an effort to keep the illustration work in-house. It was a costly endeavor, as Hevelius readily pointed out to colleagues: "I have already spent vast sums for the good of learning on splendid astronomical instruments and the publication of books."[21] The works themselves were of such high quality that Henry Oldenburg asked Hevelius to do work for the Royal Society on commission, but Hevelius declined. He did, however, request that Oldenburg use his books as a kind of currency to purchase books and materials in London that he could not access in Gdańsk. "I should send to London at my own risk enough books to be worth forty or fifty pounds sterling; as much of this sum as necessary should be given to the workman for the purchase of the best telescope and the rest reserved for buying either some books or other things I need."[22] This comment alerts us to yet another motivation for an author to self-publish: they retained the value of the copies printed, which they could then "spend" as they needed. Hevelius requested that Oldenburg distribute other copies of his works to fellows with an interest in astronomy (Hooke, Wallis, Ward, and Oldenburg) and to the libraries at Cambridge and Oxford. Finally, he asked that the remaining books be sold in London shops at a specific price. Copies of his *Cometographia*, printed

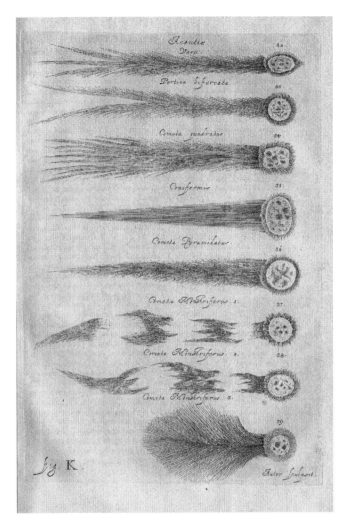

Johannes Hevelius, *Cometographia* (1668). *Courtesy of the Library of Congress.*

on thicker paper, were to be sold for 5 1/2 imperials.[23] In late 1668 when Oldenburg wrote Hevelius to acknowledge receipt of the copies, he indicated that people were eager to read the work, and subsequent correspondence suggests that Oldenburg himself was managing the sale of the books. He informed Hevelius that he had sold nineteen copies of the *Cometographia*, at twenty-eight to thirty pounds per book, as well as dozens of other works Hevelius had sent.[24] Oldenburg also relayed the Royal Society's appreciation of the treatise. Included with his letter were seven queries from Robert Hooke to Hevelius about the *Cometographia*. Three dealt directly with Hevelius' engravings. Hooke asked, for example, "In your engravings of comets, do the parts made black and shadowy stand for the bright and shining portions of the comet, and . . . do the white and shining portions of the figures correspond to the dark, opaque areas of the comet? Were the engravings you have printed sketched at the very moment that you observed the body of the comet with the telescope?"[25]

Hooke also wondered whether the halos around the comet's head appeared in the telescope in the same way they were depicted on a particular page in the work. Such questions highlight Hooke's interest in the printing and engraving processes, and in the important question of whether printed images could faithfully reproduce observed events. Hevelius' reply came swiftly. He explained his shading scheme for the comets—that darkness correlates to greater density; lighter shading equals thinner matter—and then assured Hooke, "I have presented each and every figure of the comets that I have observed myself . . . having engraved them on copper with my own hands, and sketched them in pencil upon paper *at the very instant of time when I was observing* them with the telescope, employing every care of which I was capable. So that I have paid full attention to all those variations of matters of light and shade, many times bringing the telescope to the eye."[26] Hevelius emphasizes the importance of bringing the eye, the instrument, and the hand into close proximity, reducing both space and time between observation and sketch, sketch and engraving. In her study of the images in Hevelius' *Selenographia* of 1647, Katherin Müller describes "a specific visual grammar according to which Hevelius composed both his drawings and the final prints. While the image seemingly captures nothing more than the phenomenon itself . . . it actually follows an interpretative system."[27] Müller quotes Hevelius on his technique for creating images of the moon: "I chose such a way to sketch out and overshadow that I can render sufficiently all the most different colours that can be seen on the Moon . . . Therefore I engraved the bigger spots indicating lakes with double horizontal dark lines; I surrounded the mountains and valleys with lines; the tops and peaks of the mountains are expressed by circles, certainly with the objective of rendering them better and more luminous."[28] Hevelius' attention to details of shading and light would have been welcomed by Hooke, who expressed concern over the kind of misrepresentations of nature that resulted from "Mr. Engraver's Fancy," when the engraver "knows no more the truth of things to be represented, than any other person."[29] Hooke was certainly in a position to speak on the matter, having spent a great deal of time thinking about the representation of three-dimensional objects—particularly those viewed through a microscope—in two dimensions. The engravings of his *Micrographia* reveal his desire to depict nature in its "true form." As Meghan Doherty has shown, it was Hooke's intimate knowledge of print culture and the art of engraving that facilitated this.[30] Hooke spent time in print shops and understood the tools and techniques of their trade. He collected a variety of books on printing and engraving, and his diary records the many times he offered printing advice, even to those in the business.[31]

Hevelius, like Tycho before him, sat at one end of an authorial spectrum, where an individual with resources could own and operate a full-size printing press and could take the time to learn the skills required to engrave images.[32] At the other end of the spectrum were smaller, more affordable, and nimbler printing techniques, such as the one developed by natural philosopher William Petty. Petty was a charter member of the Royal Society, known for a diverse array of interests from mathematics to medicine. He invented a twin-hulled boat for the navy—mocked by some but lauded by naval administrator Samuel Pepys—and, perhaps most importantly, applied the empirical framework of natural philosophy to economic thought and analysis.[33] He also invented a printing machine that was first described in the preface to a 1647 treatise on education. The work was dedicated to "his honoured friend Master Samuel Hartlib." He requested Hartlib's feedback on what he called "double-writing" but insisted that the idea be kept quiet. "I shall desire you to shew [my instructions] unto no more than needs you must, since they can please only those few, that are Real Friends to the Design of Realities." Petty expounded on his technology more fully in a seven-page pamphlet entitled *A Declaration Concerning the newly invented Art of Double Writing*.[34] Here he described the invention as "an instrument of small bulk and price, easily made, and very durable, whereby one may in an houres practice, write two Copies of the same thing at once . . . and in a few dayes further use thereof, recover the former fairnesse and swiftnesse of hand."[35] Petty claimed the instrument would benefit lawyers, merchants, and especially scholars, whose efforts to transcribe rare manuscripts would benefit from an easy method of duplication. He noted two versions of the implement, one for making duplicates and another suited to making more than two copies. The latter would not be readily available but could be built on demand if needed.[36] To further promote his

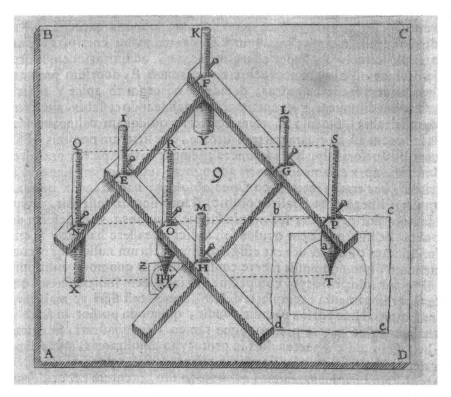

Christoph Scheiner, *Pantographice* (1631). *Courtesy of the Huntington Library.*

invention, for which he solicited buyers in subscription fashion, Petty printed a broadside describing his double-writing technique and advertising his patent, which he received in 1648 for a duration of fourteen years. The broadside was circulated and included with the instrument upon purchase as a set of instructions.[37] There is no image of Petty's invention, but descriptions indicate that it resembled a device invented by the German astronomer Christoph Scheiner and presented in his *Pantographice* of 1631.[38]

Indeed, while Petty never mentions Scheiner's work, it is difficult to imagine he was unaware of it, and it is equally likely that his instrument looked and operated a lot like Scheiner's.[39] The instrument worked by fixing multiple quills to an arm that guided their motion in parallel with the writer's hand movements. The quills were fed continually and automatically with ink, so that the writer had only to attend to their own script. On December 22, 1647, Petty demonstrated his instrument in front of nine witnesses, for whom he copied out the first chapter of St. Paul to the Hebrews. Petty produced two identical copies in fifteen minutes, faster than any scrivener could have done it.[40] Though numerous scholars expressed interest in the double-writing machine, Petty struggled to amass the capital he needed to develop the idea, and he would not divulge details about his mechanism until he had financial backing. Letters to Parliament, the London livery companies, and Charles Cavendish, a known patron of the sciences, yielded no support, and the promised delivery date for Petty's instrument came and went.[41] He even dedicated the invention to Robert Boyle, writing that "if the world embrace the use of this invention, you may, by God's ordinary providence, live to see a thing, whereof you have the very maidenhead and patronage, to be of daily and almost hourly use to most men in most countries of the whole earth, and that to perpetuity."[42] No response from Boyle survives, and over time Petty's colleagues became increasingly skeptical of his ability to produce this "engine of double writing." In August 1649 Petty submitted a document to the Council of State offering to expose the workings of his machine in exchange for £1,500, but the request came to naught and the idea continued to stall. In a letter to Hartlib in 1658— a full decade after Petty had introduced double-writing—Oldenburg wrote, "I am told here, that Dr. Petty hath a commodious way of printing, called *Instrumentum Petti*, very convenient to carry about, and to print in traveling some few [sheets] of paper, if occasion present itself. I entreat you, if you have or know it, to communicate it unto me."[43] This instrument differed from Petty's earlier invention in that it did not attempt to duplicate handwriting but rather acted as a mini–printing press. Petty boasted that it would allow one to print in Chinese or Arabic, languages for which most printers did not own fonts. Samuel Hartlib noted that with this new technique printers "needeth not discompose at all. And could compose as fast as one could write faire. And print of far faster than in the ordinarie way even many copies at once. It needeth not so long a learning as the common ones."[44] In the end, Petty failed to sell this instrument—at least in any mass-produced way—but the attempt tells us as much as the results. Scientists sought a way to control the printing process, thereby, as Petty puts it, "saving the labor of examination, discovering and preventing falsification, and performing the whole business of writing, as with ease and speed, so with privacy also."[45]

Because of its potential as a reproduction technology, double-writing was taken up by several of Petty's contemporaries, including the astronomer, anatomist, and architect Christopher Wren. Shortly after Petty started his work,

Wren began developing his own method of personal printing, seemingly unaware that others had attempted it. Over the course of his illustrious career, Wren avoided traditional publication, preferring to share his findings privately through letters and exchanged manuscripts. This approach, however, belied an abiding interest in the reproduction of texts and images, a passion that led Wren to develop and experiment with numerous printing techniques. The first of his inventions was, like Petty's, a method of double-writing. He devised it sometime in the early 1650s, and it was described as "making two several pens on two several papers to write one and the same ducture of letters, with as near as possible the same beauty and facility that is found in common writing, by an instrument called the *Diplographical Instrument*."[46] Working like Petty's pantograph, Wren's device was intended to "copy out in every punctilio the exact resemblance, or rather the very identity of the two copies, as if one should fancy such a piece of magick as should make the same thing really two."[47] Wren showed his design to several other natural philosophers, who were enthusiastic about it. Among his supporters was John Wilkins, a key member of the group of scientists who gathered at Oxford. But Wren neglected to see the invention through to production. He regretted this years later, when he falsely believed that someone else had claimed credit for a double-writing implement. "I am apt to believe from good information, that those who now boast of it had it from one who, having fully seen the author's and examined it carefully (as it is easy to carry away being of no complicate composure) described it justly to his friend and assisted him to make it . . . Indeed though I care not for having a successor in invention yet it behooves me to vindicate myself from the aspersion of having a predecessor."[48] By 1654 the London instrument maker Ralph Greatorex stepped in where Wren left off, building around twelve models of the double-writing machine based on Wren's design and selling them for twelve shillings each.[49]

Wren's motivations for developing a double-writing instrument stemmed from his curiosity about mechanics and his commitment to developing technologies that would prove useful to society. An inventory of the inventions he worked on over the years bears this out. On the list, which includes artificial eyes, balances, and musical instruments, one finds the following entries: "to write double by an instrument," "several new ways of graving and etching," and "new ways of printing."[50] These entries reflect not only a broad interest in printing techniques but also Wren's curiosity about effective ways to produce images. The former he approached through his writing instrument, but the latter took him into the world of drawing, engraving, and etching, crafts that proved essential to the creation of accurate images. In his *History of the Royal Society*, Thomas Sprat specifically mentions Wren's "curious and exceeding speedy way of Etching," and Robert Hooke, in the preface to *Micrographia*, credits Wren, "whose original draughts do now make one of the ornaments of that great collection of rarities in the King's Closet."[51] Hooke notes that although he ultimately compiled the many observations that comprise the *Micrographia*, he was picking up the project only on Wren's suggestion, since the latter had moved on to other things. I will return to his other projects shortly, but first, we should consider the work of Wren's peers in the area of printed images.

Within the Royal Society there were at least a half-dozen fellows who took up the practice of engraving or etching to illustrate their own works. The earliest scientific images, of course, were created from woodblocks, particularly in the field of botany. In this relief method, the illustration was cut into blocks of prepared wood that were dense enough to maintain their integrity under repeated impressions, such as pearwood or boxwood.[52] In cases where left-right orientation mattered, such as maps, the woodblock would be cut in reverse, ensuring that the printed image was correct. Ink was applied to the raised areas of the woodblock, thus transferring the image to the page. It required tremendous skill to accurately carve woodblocks, but even with an experienced craftsman behind the chisel, shading and nuance were difficult to achieve. Thus, by 1600 woodblock printing was on the decline, to be replaced by engraving.[53] An intaglio method, engraving was a technique wherein grooves were made on the surface medium—typically a copper plate—so that ink was applied to the plate and the plate subsequently wiped. The ink that remained in the grooves would be transferred to the paper. Because the copper plate was supple, the engraving tool called a burin or graver could create a range of deep and shallow lines, resulting in images with tremendous subtlety. An experienced engraver could achieve detailed gradations of light and dark.

With its name derived from the German word "to eat," etching was similar to engraving, but rather than using a burin to cut into the metal place, acid would eat away the metal to create a recession. The craftsman would receive a plate covered in an acid-resistant substance (called a ground), such as wax. They would then use a tool to draw through this layer, exposing the plate below. When the entire thing was covered in acid, only the exposed areas would be eaten away by the acid, thereby creating grooves, just as in engraving. Ink is then applied to the plate, wiped off the surface, and held in the grooves until an impression is made. Gradations in shading can be achieved through re-peated acid baths. Compared to engraving, etching is a much easier process since the craftsman need not apply pressure to the plate itself in creating grooves. Etching works more like typical drawing and is more forgiving to the amateur draftsman.[54]

John Evelyn, Fellow of the Royal Society and printing enthusiast, was keen to highlight the quality of the engraved images in the work of numerous sci-entists, many of whom had done the work themselves. In his 1662 treatise *Sculptura: The history and art of chalcography*, Evelyn offers an expansive his-tory of European printmaking and details the importance of such techniques as etching, engraving, and the newer art of mezzotint. The latter was perhaps the most exciting, as it offered the prospect of creating illustrations—botanical, anatomical, astronomical—that were tonally complex with subtle gradations of shading so that, Evelyn wrote, "a print [could] emulate even the best of drawings."[55] He praises the astronomical engravings done by Hevelius, which we encountered above. "The learned Hevelius has shewed his admirable dex-terity in this art . . . and is therefore one of the noblest instances of the ex-traordinary use of this talent for men of letters, and that would be accurate in the Diagrams which they publish in their works."[56] Likewise, he lauds the work of sixteenth-century mathematician John Blagrave, whose treatise *The Math-ematical Jewel* includes numerous fine illustrations of mathematical instru-ments. Blagrave's frontispiece notes that he "cut all the prints or pictures of the whole work with his owne hands."[57] Evelyn continues on to highlight typo-graphic exemplars from the period, from Albrecht Dürer to Christopher Wren.

Evelyn's appreciation for the art of engraving was tied to his broader phil-osophical desire to link science and the trades. In the 1650s, before the Royal Society was even established, Evelyn had compiled observations for a treatise entitled *Trades: Secrets & Receipts Mechanical*.[58] In the spirit of Francis Bacon, this work was designed to consolidate practical knowledge for the benefit of the commonwealth, and it included practices related to printing and engrav-ing. By the end of the decade, however, the work had stalled, as Evelyn tried to negotiate the boundary between craft and academic knowledge. In a letter to Robert Boyle he expresses fears that publication of such information might compromise the craftsmen whose livelihoods relied on secrecy. Publishing would "debase much of their esteem by prostituting them to the vulgar."[59] He goes on to explain, "I conceived that a true and ingenious discovery of these and the like arts, would, to better purpose, be compiled for the use of that Mathematico-Chymico-Mechanical School . . . where they might (not without a note of secrecy) be taught to those that either affected or desired any of them: and from thence, as from another Solomon's house, so much of them only made public, as should from time to time be judged convenient."[60] Evelyn's main concern, then, was that the knowledge be made available only to the virtuosi who could be trusted to use it to improve and disseminate their research. This sentiment was echoed in *Sculptura* twelve years later. Ultimately, however, Evelyn refused to disclose the process for creating a mezzotint print, claiming, "I did not think it necessary that an art so curious and (as yet) so little vulgar . . . was to be prostituted at so cheap a rate, as the more naked describing of it here, would too soon have exposed it to."[61] Thus, as had happened with his treatise on mechanical trades, Evelyn avoided full disclosure of his printmaking technique, limiting access to those gentlemen of science whom he considered worthy and who could acquire it from him personally.[62]

On June 11, 1662, Evelyn presented his *Sculptura* to the Royal Society, a body that would officially receive its charter from Charles II just a month later. In the book's dedication to Robert Boyle, Evelyn expresses hope that gentlemen of the nation would find his work useful, "and, especially, that such as are addicted to the more noble mathematical sciences, may draw, and engrave their schemes with delight and assurance."[63] For Evelyn, it was precisely those involved in the scientific disciplines who would benefit from his work. "It is hardly to be imagined of how great use, and conducible, a competent address in this art of drawing and designing is to the several advantages which occur; and especially to the more noble mathematical sciences, as we have already instanced in the lunary works of Hevelius, and are no less obliged to celebrate some of our own countrymen famous for their dexterity in this incomparable art."[64] Evelyn understood that nature could be described in words and equations, but for him nothing captured the essence of the natural world as efficiently as images. "Words are never so express as figures."[65] Echoing this sentiment a decade later is William Salmon's *Polygraphice*, a comprehensive synthesis of other people's treatises related to engraving, etching, drawing, and other arts.[66] Salmon, whose prefaces we encountered in a previous chapter, came from a medical background and thus viewed drawing and the print arts as intimately related to understanding the body. He devotes his first chapter of *Polygraphice* to the drawing of faces, extremities, and whole bodies, emphasizing proportionality and shadowing. He also explains methods of duplicating images by grinding printer's ink, diluting it in water, and then using the ink to trace an outline of the desired image. Wet paper is then placed on top of the ink to transfer the image. His second chapter addresses the tools and techniques of engraving, etching, and limning, representing a thing in a painting or drawing. Several sections on perspective also summarize the techniques and shortcuts in creating depth of field. As scholars have shown, Salmon drew heavily on other scholars' writings, so we cannot consider his contributions original, but the fact that *Polygraphice* went through eight editions is testimony to its popularity.[67] It is also a reminder of the importance of printing technologies to the medical sciences.

Like William Salmon and John Evelyn, Christopher Wren was deeply engaged with the craft of producing images. In *Sculptura* Evelyn characterizes Wren as "that rare and early prodigy of universal science" whose engraving work was notable. And when John Aubrey sought an engraver to produce images for his natural history of Britain, his friend John Hoskins recommended "Dr. W [Wren] or Mr. Hooke to teach a new short cheap way of etching, else the book will be too dear."[68] Wren's skills, refined through extensive experimentation and practice, are nowhere more evident than in his eight etchings for Dr. Thomas Willis's famous anatomical study of the brain, *Cerebri Anatome*, published in 1664.

Nathan Flis, who works on seventeenth century British art, has argued that "Wren's figures set a precedent in English works of natural philosophy and natural history, which was immediately followed by Hooke's large-scale etchings for *Micrographia*."[69] Willis had his own praise for Wren in the book's preface, noting that he had done the figures of the brain and skull with great exactitude. Wren's efforts in the world of printing reflected attention to both text and image, though as far as we know he did not combine them in a unified printing technique. That task would be taken up by some of Wren's colleagues, starting with Christiaan Huygens.

As we have seen, Huygens had intimate knowledge of print culture. In much the same way that he cultivated a technical understanding of lens making and the construction of clocks, he became acquainted with the work that transpired inside the printing house, and he developed his own technique for printing. On May 19, 1669, he wrote to Henry Oldenburg, "To give you new invention for new invention, I send you a sample of my new printing process in the leaf you see here. It is intended for printing writing and also for geometrical

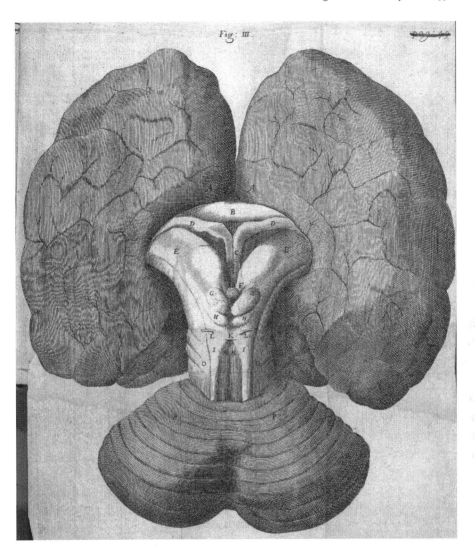

Thomas Willis, *Cerebri Anatome* (1664). *Courtesy of the Huntington Library.*

figures. It is cheap, and can quickly both engrave plates and print. Your Fellows will not find much difficulty in guessing how it is done; otherwise I shall provide an explanation if they wish."[70] With his new method, Huygens printed a quotation from Virgil, but instead of setting the quotation, his "printed" material appeared in his own handwriting, almost as if it were an etching. In addition to sending it to the Royal Society, he sent a copy of the print to his brother Lodewijck, inviting him to deduce the method from the sheet.[71] Several weeks later Oldenburg responded that several fellows of the Royal Society had proffered a guess about the technique, and some claimed to have a similar method. They agreed to send details of their inventions to Huygens. According to Oldenburg:

> By one of our methods, Sir [William] Petty's, as many copies as one chooses may be printed while the book is on sale; and after an edition has been sold out it is possible to print a second, a third, and so on to any desired number . . . I believe that Mr. Wren has another method which is perhaps similar to yours . . . I do not know whether your method will print as many copies as one wishes, whether one can print with it matter that is printed already, and whether ordinary printer's ink is used. You will, if you please, inform us on these points.[72]

It is interesting that Oldenburg describes Petty's instrument as one that can generate more than two copies, since Petty described his invention as one focused on making a single copy rather than multiples: "As for the Instruments of multiple writing I would not have them expected, because their use is not generally needful, and incomparably more hard then [sic] of those for double only: Wherefore I have thought it rather fit to reserve them for the peculiar use of some."[73] It is likely that, writing his letter to Huygens more than twenty years after Petty first published on his instrument, Oldenburg simply described the piece inaccurately. Regardless, his letter implies that Petty's was an early and viable technique, one of several that had been developed by natural philosophers around London. In his subsequent reply, Huygens agreed that Petty's method sounded similar to his own, but he continued to withhold details of his printing method until others had the chance to study it. He also offered a newly printed work for examination: "I wanted to try to print geometric figures also by this method; this succeeded only fairly well as you will see by the example I send you, containing the construction for a problem which I solved lately and which our mathematicians thought pretty elegant."[74] This printed sheet appears, like the Virgil quotation, in Huygens' hand rather than set in type; nevertheless, it could be reproduced as often as necessary. In fact, the geometric diagram from this sheet was reproduced in the *Philosophical Transactions* article on Alhazen's problem in October 1673.[75] On July 5, 1669, Oldenburg sent Huygens the Royal Society's best guess for his printing method:

I have shown our Fellows the geometrical figure which you have had printed by your new method. They ordered me to return you thanks, and Mr. Wren conjectures that you use the following method: Taking a brass plate as thin as paper you cover it with a varnish suitable for engraving and have the design drawn on that (taking care not to close up the letters) with such strong nitric acid as quite to pierce the brass. When this is done you turn the plate, putting it on another which is thicker and entirely coated with printer's ink; and then you pass it through a rolling press in the usual way. You will please tell us if Mr. Wren has described this correctly or not. Sir William Petty's method is different; but as he is not in England I have not permission to reveal it at present.[76]

Huygens' account of his new method, recorded among his papers, reveals how close Wren's guess was. He did use a brass plate, which he covered with a layer of wax, a setup similar to that used in etching. Then, he explains, "it is written on with an iron point, and is then plunged into acid which *wholly eats away the letters*, leaving them open. The wax is then melted away with a flame."[77] In essence, Huygens used the acid to cut into the exposed parts of the plate, creating a kind of stencil. He then applied a layer of printer's ink to another, thicker plate, and then stacked the first plate (which held the etching) on top of it, so that the ink could come through only where the acid had eaten away the words. He continues, "A damp sheet of paper is placed on the thinner plate, soft leather is placed over this, and then a thicker cloth. The whole is then subjected to pressure in the copperplate engravers' press that consists of two cylinders." What resulted was a copy of the original etching printed on the paper. After he completed several printing trials, Huygens made additional notes to himself about improvements he hoped to make to his method. These included reducing the viscosity of the printer's ink, ideas for layering the plates and cloth in such a way as to reduce the pressure of the press, and observations on the difficulty forming letters that were similar in appearance, such as "a" and "o," that could be hard to differentiate once printed.[78] Among Huygens' papers, housed at Leiden University, there are numerous artifacts related to his experiments with printing and engraving. These include copies of poems and phrases, mathematical samples, and even the original brass and copper plates he used. It is noteworthy that the basic material Huygens used to develop his printing method, metal plates and wax, was precisely the material he had worked with seven years earlier when building an air pump. This is not to say that his air pump design had any direct impact on his print-

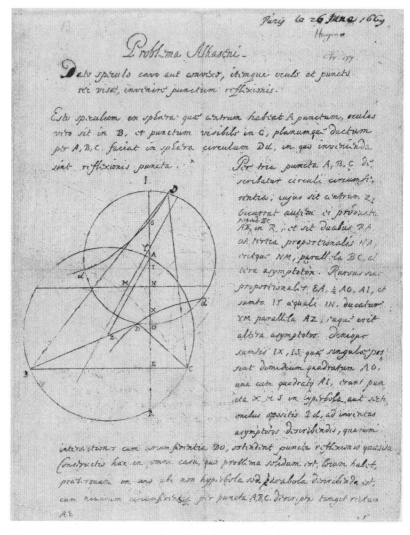

Christiaan Huygens, *Printing Sample of Al-Hazen's Problem* (1669). *Courtesy of the Royal Society.*

ing method but rather to reinforce that Huygens was both adept at instrument building and extremely familiar with the materials required for his print technology.

The individuals I have examined so far demonstrated a significant understanding of printing and engraving techniques. We should not consider their familiarity with these skills unusual. The exchanges between mathematicians John Collins and James Gregory are illustrative of authorial awareness of the technical side of print. Collins, who earned a reputation for helping others get their work into print, urged Gregory to publish his mathematics and offered his own help in overseeing the press: "I much rejoice at your inclination to publish your lucubrations and willingly offer my best assistance in correcting at the press, &c."[79] More will be said in the following chapter about this kind of editorial encouragement and assistance, but what is telling here is Collins' deep understanding of the printing process and the specific issues that arise in the publication of mathematical texts. First, Collins points out to Gregory that the few printers in London capable of doing mathematical works already had a backlog of jobs and were unlikely to take on new projects. The only printer who had the requisite skills and sorts (pieces of type) to print mathematics— William Godbid—had at least six works in his queue, each of which Collins readily named. He suggests Gregory consider a press in his home country of Scotland, and in the event that there was no one available to do the woodcuts there, Collins offered his assistance, "to new draw the schemes and get the plates graved here."[80] He goes on to focus on small printing details that would improve Gregory's work in the press: "As to your book give me leave to suggest a few things to prevent mistakes, first that some persons make the letter b and figure 6 so like, that the one may be mistaken for the other, not to use the letter e which if worn prints black, and may be taken for c as in Dr. Wallis his works. Pell's method of putting a great letter for a known quantity, and a small one for an unknown is much approved."[81] Collins discusses such formalisms as the use of italic in equations, and he encourages Gregory to include some applied examples in his treatise for didactic purposes. Through his various letters, we see in Collins not only an adept mathematician but one whose familiarity with the world of print proved to be a great advantage both to him and his colleagues.

The examples provided above illustrate the deep and widespread interest of early modern scientists in printing technologies. Taken collectively, however, they hardly revolutionized the contemporary printing paradigm. Do-it-yourself printing, etching, and engraving never replaced the work of printing processes and commercial publishing. The few cases in which an author had the financial resources to construct and operate their own full-scale presses are exceptions to the rule. Nevertheless, the effort put into developing new printing and engraving techniques is a potent reminder of the status of the press as a scientific instrument, one that demanded refinement and customization. Joseph Moxon, the famous London printer, provides the definitive example of the way the printing press could be framed as something more than an artisanal tool. It had the potential, in his estimation, to be understood as a kind of mathematical instrument in its own right. In his famous treatise on typography, *Mechanick Exercises on the Whole Art of Printing* (1683), Moxon presented the first comprehensive study of printing in the West, and though we might instinctively consider him a tradesman, he was someone who had a clear hand in the applied sciences. His press issued works of astronomy, mathematics, instrumentation, and geography; it is no surprise that he was made a fellow of the Royal Society in 1678. With status as a fellow providing legitimacy, Moxon argued that typography should be considered one of the mathematical arts. In framing this assertion, Moxon drew heavily on the work of John Dee, the English mathematician who, in his famous preface to Euclid's *Elements*, drew on classical examples to argue for architecture's place among the mathematical arts. Moxon appropriated Dee's logic to insist that "a typographer ought to be equally qualified with all the sciences that becomes an architect." Lest the reader question his syllogism, Moxon elaborates on his conception of typography: "By a Typographer, I do not mean a printer, as he is vulgarly accounted, any more than Dr. Dee means a carpenter or mason to be an architect . . . I mean such a one, who by his own judgment, from solid reasoning with himself, can either perform, or direct others to perform, from the beginning to the end, all the handy-works and physical operations relating to typographie."[82] Matthew Hunter has unpacked Moxon's theory of typography and examined the links between his ideas and those of Dee: "Architect here denotes the function by which all practical crafts and bodily skills (in Moxon's case, making and composing type, imposing and printing forms, proofreading, and binding) become integrated under the governance of a theoretically informed intelligence."[83] For Moxon, the broad spectrum of tasks that made up typography, and which he admits have been divided up within the print shop, can be considered a "mathematical art" when taken together. This perspective is integral to the way early modern scientists understood the press. Their efforts to reimagine printing techniques were tethered to their view of the press as a scientific instrument, a tool of mathematics, mechanics, and medicine that, like the air pump or barometer, could yield authoritative products of scientific thought. John Evelyn, Christopher Wren, Christian Huygens, and the many others who had a hand in printing engaged with the press in this spirit. They viewed this instrument much like they did the others that passed through their studies, as a tool for creating knowledge, both textual and visual, that was imbued with authority. When they developed printing methods to reproduce their own work, they not only wrested control away from a commercial endeavor that could compromise their efforts; they proactively shaped the final product in a way they believed would most benefit their ideas.

Chapter Five

Silent Midwives

The Role of Editors in Early Modern Science

The circling streams, once thought but pools, of blood,
(Whether life's fuel, or the body's food)
From dark oblivion Harvey's name shall save
While Ent keeps all the honour that he gave . . .
Such is the healing virtue of your pen,
To perfect cures on books, as well as men.   —JOHN DRYDEN

Twenty-five years ago, Robert Darnton posited a communications circuit that traced the agents involved in the production of texts, from author to printer to booksellers and finally to the audience, who in turn influenced authorial decisions about publishing.[1] The staying power of this model, which Darnton intended as a heuristic tool, can be attributed in part to the fact that it encouraged historians to think of the history of books holistically. His circuit is a reminder that individual decisions and actions are rarely well understood in silos, and that the publishing world was—from its inception—a terribly complex and interconnected place. The importance of Darnton's foundational schema is seen in the waves of subsequent research it inspired, some of which teased out fascinating stories about piracy, about the intrigues and culture of the print shop, and about readers' reception of works. The circuit described by Thomas Adams and Nicholas Barker, for example, looks not only at the constituents involved in print but also at the socio-economic pressures that influence the publication, distribution, and reception of works.[2] Their model effectively inverts Darnton's, emphasizing the book itself, rather than the people who produced it. This chapter takes its cue from both schemes by considering social and intellectual forces (commercial pressures, cultural expectations, religious contexts) as they influence a character who—at least nominally—was not part of Darnton's scheme: the editor. He did reference the publisher, which in seventeenth century terminology could equate to the role of editor. The Latin term *editor*—stemming from the verb *edere*—is defined as the producer or publisher of a thing, and the *Oxford English Dictionary* puts its first use in 1649. Only in the early eighteenth century (1712) was its meaning expanded to include the work of selecting, revising, and arranging material for publication. But this is a case where the vocabulary had simply not caught up with the activities of those involved in print. There were clearly individuals in the seventeenth century who played an intermediary role in the production of printed works but who were not technically publishers. Into Darnton's circuit, therefore, I have added the editor.

In what follows I examine the role of editors in the production of scientific treatises, considering both renowned and lesser-known examples of editorial influence that shaped, among others, the anatomical work of William Harvey, the mathematical and astronomical publications of Christiaan Huygens, and the natural philosophical work of Robert Boyle. Covering several disciplines in the sciences, these cases highlight the array of tasks an editor would assume and the many ways their influence can be seen on the final, published work. The work of these individuals—like that of amanuenses or laboratory assistants—is often sidelined or outright ignored, while historical attention turns to the authors themselves, whose names and theories have risen to the surface of the historical conversation. Like Steven Shapin's "invisible technicians," they operated in the background, doing a tremendous amount of detail work on scientific manuscripts. Yet even the people who employed them lacked a nomenclature to describe their function, and their status in the world of print was hardly codified. Sometimes called "correctors," other times "publishers" or "assistants," the individuals who shepherded works from the author's study to the compositor's stick to the booksellers' stalls largely disappeared once a work was published.[3]

To get an understanding for the tasks involved in printing, some of which editors would take on, we can look to two contemporary works on topic: Hieronymus Hornschuch's *Orthotypographia*, published in Leipzig in 1608, and Joseph Moxon's *Mechanick Exercises*, published in London in 1683. Hornschuch's treatise presents one of the earliest detailed descriptions of the editorial work that took place inside the print shop. Specifically, he focused on the task of the corrector, a job he held for ten years while a student. The corrector compared a printed proof with the original holograph, which was too often in a difficult or messy hand, ensuring that they matched as closely as possible.[4] The corrector marked any discrepancies so that the compositor could come

OFFICINÆ TYPOGRA-
PHICÆ DELINEATIO.

E N Thymii fculptoris opus,quo prodidit unâ
Singula chalcographi munera ritè gregis.
Et correctorum curas,operasq; regentum,
Quasq; gerit lector, compofitorq; vices.
Ut vulgus fileam. tu qui legis ifta, libello
Fac iteratâ animi fedulitate fatis.
Sic meritæ cumulans hinc fertilitatis honores,
Ceu pictura oculos, intima mentis ages.

L. I. L. F.

Hieronymus Hornschuch, *Orthotypographia* (1608). *Courtesy of the
Hanna Holborn Gray Special Collections Research Center, University of
Chicago Library.*

A   Si quid eſt omiſſum, litera ſit, ſive dictio, ſive diſtinctionum aliqua, in locum illius omiſſi hoc ſignum
in linea inſeritur, & quod omiſſum eſt, ad marginem è regione verſus dextram adſcribitur. Vt ſi e. g.
Imprator poſitum ſit pro Imperator, ibi inter p & r
dictam notam interpono, & e literam ad marginem
annoto.

§   Quod abundat & ſuperfluum eſt, illud calamo in-
ductum hæc in margine poſita expungere iubet.

ſ   Cum litera aut nota diſtinctionis eſt inverſa, can-
cellatur, & hoc in margine ſignum adponitur.

   Si quando dictiones, quæ ſciungendæ erant, omiſſo
ſpacio iunguntur, hac nota divelluntur, & ipſa ad
marginem quoq, ponitur.

   Divulſa & hiantia coniungit, eo quo in præce-
dente dictum, modo adhibita.

   Quæ diſtorta ſunt & oblique poſita; aut quæ al-
tius exſtant, quò vicina litera minùs exprimantur,
hoc ſigno intus & in margine notantur.

1 | 2   In μετ... dictionum ad marginem ponitur, &
dictiones, quo reponendæ ſint ordine, ſupra numeris;
infrà lineá ſubductá notantur.

   Spacium deprimendum ſignificat.

⊙   Puncti nota.

(?)   Omne diſtinctionum genus, ut ſunt Colon, Semi-
colon, Interrogatio, Exclamatio, Admiratio, &c.
ſemicirculo, quali hic Comma, ad marginem poſito,
inſcribitur.

(–   Nota diviſionis, in extremitate lineæ, quæ integram
aliquam dictionem capere non potuit.   Si

Hieronymus Hornschuch, *Orthotypographia* (1608). *Courtesy of the Hanna Holborn Gray Special Collections Research Center, University of Chicago Library.*

back and make changes in the type. To facilitate this, Hornschuch developed a list of proofreading marks to make correction more efficient.

These symbols would eventually become standardized among copy editors. Most importantly, Hornschuch's manual professionalized the corrector's role. The engraved image of a print shop found in his book includes the corrector, who sits in the back right corner of the shop with an ink pot, marking up the proof while the author stands by, gesticulating in an animated fashion.

In Hornschuch's experience, it was critical to have trained correctors—either inside the print shop or hired from outside—to review proofs, rather than having an author edit their own work. On this point he is firm:

> Thus in connection with what has been said before, if good seed has been scattered in the typographical field, it will be a hard job for wretched weedy errors to grow there, and there will be no need for anyone who lacks trust in the corrector to read his own proofs and to purge them of such hated tares—a practice which I can testify has been generally harmful to all such people, as is shown by the fact that never were more mistakes found afterwards than in their own published works. For it is impossible for anyone not fully trained in this school not to go wrong. Therefore it will be best to entrust the work to the corrector who is in duty bound to check and examine everything two or three times.[5]

Hornschuch's treatise was a plea for authorial trust in, and deference to, the correctors. He claims he had encountered more than two thousand errors in a printed proof, most of which he attributed to a poor copy from the author. "In the face of faulty manuscripts, which cannot be read except with extreme difficulty," Hornschuch writes, compositors are helpless. When they end up producing error-laden copy, "the punishment is meted out in the end to the corrector, who is distracted and almost worn out in expunging so many errors at his own cost and the expensive outlay of his time."[6] Only with a good corrector could the errata be contained.[7]

More than seventy years after Hornschuch, Joseph Moxon continued the tradition of offering technical manuals for printers in his widely used *Mechanick Exercises*. This was the first English work written for tradesmen about the craft of printing. Moxon's detailed description of the entire process, from the job of letter cutting to the role of the compositor to the pressman's trade, reminds us of the level of technical expertise required and how many hands were involved in printing a work. Nevertheless, Moxon's exhaustive book does not use the word "editor," nor does he speak explicitly of the process of editing. Like Hornschuch he refers only to correcting, a task he says can be done either by the compositor at the front end, working between the author's handwritten copy and the type they are setting, or else (and more likely) by a trained corrector at the back end, working from a printed proof. Correctors might also perform "casting-off," wherein the length of a manuscript was converted into an approximate length for the printed text. Once this was known, a work could be divided up by signatures—groups of pages printed together on a single sheet—and delegated to different employees of the print shop. The efficiency this process afforded was hardly trivial, as the ability to print pages out of order, as opposed to serially, allowed for a more rapid production of the finished product. As the first readers of a text, correctors had tremendous influence on what would become the final product. Their role was being quickly being formalized. By the time the popular *Printer's Grammar* (1755) was published, the labor of correction was clearly distilled: "The proof being now read, and the real faults marked distinctly and fair, the corrector examines the pages of the sheet, or form, whether they are imposed right . . . After which the proof is given to the compositor, to correct it in the metal."[8] If resources permitted, a print shop might hire someone to do correction work, perhaps even a "learned corrector" with academic skills. The work of such individuals was considered to be "more profound and more personal" than others.[9] Here Moxon and his fellow printers had in mind the scholars who corrected in the large printing houses of France and the Low Countries, where correctors were versed in many languages and familiar with the classics.[10] Christophe Plantin, for example, typically hired one corrector for two to three presses.[11] Conversely, smaller printing houses relied on the master printer or a skilled journeyman to handle correction, assuming an author did not have the means to hire someone themselves. In addition to checking the fidelity of the proof to the original copy, a corrector would check signatures and page numbers to confirm that they were in the proper order.

This review of the corrector's role serves as a reminder of the kind of editorial work that existed in the early years of print. Correcting generally took place inside the print shop, but there are examples of printed proofs being sent to an author or assistant outside the shop for correction, however much the professionals wished it otherwise. We will examine specific cases of this below, where scientific texts are concerned. Often these external efforts were made by individuals who were doing more than correcting, and thus, the lexicon of "correction" is insufficient. For them, reviewing proofs was just one of a series of tasks required to see a work into print on behalf of an author. A more expansive understanding the early modern corrector-turned-editor comes from the humanist tradition of textual criticism and the demand for production of new editions of classical works. Brian Richardson's study of Renaissance print culture traces the emergence of the editorial role in the wake of the press's invention and focuses on Italian printers who hired "editors" to prepare a work for the press. In this context correcting meant more than comparing printed proofs with the author's original copy. It implied stripping away layers of added text and commentary in an effort to restore classical works to their original state. One historian describes their tasks as varieties of correction: *emendare, corrigere, emaculare* (emending, correcting, cleaning), and as early as the sixteenth century authors distinguished between correctors and proofreaders.[12] Correctors would update the translation, fix punctuation, augment the text with images, or add supplemental information that would assist the reader.[13] Despite the significance of these contributions, correctors are scarcely mentioned in the texts themselves. If they appear, they are identified only in a colophon or a passing dedicatory compliment. By the early sixteenth century, however, printing houses were competing with one another to produce the best editions of classical works. Such competition made it advantageous to identify those individuals who helped shape a work—correctors, translators, editors—and to highlight their credentials as a means of promoting it. In 1500, Venetian printer Aldus Manutius formed the "New Academy" of Hellenists, men whose task it was to select and edit a new Greek work for publication each month.[14] The Greek editions published by Aldus were marketed based on the expertise of the editor, whose name began to appear on the title page alongside that of the author and printer.[15] Scholars could presumably trust that they were getting the best possible version, one that read as closely as possibly to the original in form and meaning. They would even benefit from an editor's commentary or those "exegetic items" inserted to clarify a work's meaning.[16]

Some authors—such as Erasmus—acted as their own "learned correctors" when they sought to ensure the fidelity of the printed work, going so far as to move into the printer's house—literally above the print shop—to be on site to oversee the printing.[17] These men were sometimes referred to as "author-editors," and their intimate interaction with the proofs extended beyond what most correctors did as part of a routine job.[18] Not only did Erasmus live in Venice, near the print shop of Aldus Manutius, for nine months, he later resided with, and worked for, Johann Froben in Basel. There he did editorial work for Froben on a range of texts, including his own, and he oversaw the work of the compositors and corrected proofs while his New Testament and translations of the church fathers were printed.[19]

I have reviewed the role of learned correctors, and the various tasks they took on, to better contextualize the work of scientific editors in the seventeenth century. Practices that became standardized in the printing of classical, legal, or theological works were naturally brought to bear on the publication of scientific treatises. And as I have demonstrated in previous chapters, print culture presented numerous challenges to natural philosophers and their colleagues in ancillary disciplines. Controversy was a bitter consequence of some publishing efforts, and authors sought to avoid it where they could. There was also the tedium of preparing a manuscript for the press and making arrangements with printers—tasks many authors were happy to delegate to a willing peer. Complicating things further, there was the financial risk of publishing that someone—the author, an editor, or the printer—had to assume. And finally, authors worried that their work would not be received in the way that they hoped. In the face of these and other concerns, it is understandable that scientific authors turned to editors who could assist. Much more than correctors, these individuals acted as midwives—a term they invoked more than once—to the birth of their colleagues' work. The influence of editors on many canonical works in science was significant, and in a few cases their efforts have been well documented.[20] More generally, however, the editor's work on behalf of early modern science has been sidelined, their names relegated to a single sentence in a prefatory epistle. In what follows I want to examine specific cases in different scientific fields where editorial intervention proved instrumental to the publication of texts, and I want to understand the role of these editors as it relates to authorial attitudes toward print culture.

William Harvey and George Ent

In 1628, the Englishman William Harvey published *On the motion of the heart and blood*, wherein he proposed the radical theory that blood circulates in the body, traveling out through arteries and back to the heart through veins and pulmonary circuit.[21] This major breakthrough in physiology was hardly embraced by the medical community, which was entrenched in traditional Galenic ideas of the body. Even proponents of empirical medicine balked at Harvey's so-called ocular demonstrations, some rejecting his claims based on their dislike of vivisection as an anatomical tool. John Aubrey, a contemporary and friend of Harvey's, recorded the general response to his work: "I have heard him say that after his booke of the Circulation of the Blood came-out, that he fell mightily in his practize, and twas believed by the vulgar that he was crack-brained: and all physitians were against his opinion, and envied him; many wrote against him."[22] Among those who attacked the new anatomical model was French anatomist Jean Riolan (1577–1657), professor and later dean of the Faculty of Medicine in Paris. In 1648 Riolan published an anatomical work that argued against Harveian physiology. He believed blood did not circulate through the body nearly as often as Harvey proposed, and he thought that blood made the heart beat, rather than the reverse—the heart impelling the blood. Harvey was not inclined to respond to counterarguments, particularly when they were not supported by empirical evidence. "Perish my thoughts if they are empty and my experiments if they are wrong . . . [I]f I am wrong, let my writings lie neglected."[23] He had no desire to "trouble the Republic of Letters" on his own behalf.[24] Harvey's colleague Dr. George Ent, however, had no such qualms and stepped in as a willing and potent defender Harveian physiology.[25] Like Harvey, Ent had earned his medical degree in 1636 from the University of Padua, then the most prestigious institute for medical study on the Continent. Three years later he became a fellow of the Royal College of Physicians, the governing body of medical practice in London. A well-received anatomy lecture, delivered before Charles II in 1665, resulted in Ent's knighting at the subsequent reception, a mark of his standing in the college.[26] He was also an original member of the Royal Society, where he actively engaged in anatomical experiments with other fellows.

Ent's most significant contributions to medicine came through his efforts to support Harvey's work, which he did in two ways. First, he publicly defended Harveian circulation in the 1641 *A Defense of the Circulation of Blood*. The treatise was addressed to Emilio Parigiano (1567–1643), a physician from Venice who joined the anti-circulation faction in the wake of Harvey's discovery. Parigiano had attended medical school in Padua at the same time as Harvey and Ent, but he rejected Harvey's theory from the outset. He first refuted circulation in print in 1635 and then published a second edition of the rebuttal in 1639. The latter publication included a reprint of Harvey's original treatise. Word of Parigiano's publication spread rapidly. In February 1639, Descartes learned from Mersenne of "an Italian medical man [who] has written against

Harvey's On the Motion of the Heart."[27] Ent was aware of the debate and interceded. He responded to Parigiano point by point, refuting the Italian doctor's logic and, through a series of digressions, reasserting the anatomical basis for circulation.[28] This effort on Harvey's behalf frames the intellectual and personal relationship the two men shared leading up to their collaboration as author and editor.

In addition to defending Harvey directly, Ent assumed an editorial role in order to ensure that Harvey continued to publish despite his tumultuous experience with the circulation work.[29] In the wake of De Motu Cordis, Harvey turned his attention to various lines of anatomical experimentation, including dissections and vivisections that would reveal the mechanisms behind reproduction. In the winter of 1647, Ent visited Harvey outside of London and found him deeply engaged in numerous anatomical studies. As Harvey described his work, Ent claims to have interrupted him and explained that "many learned and judicious men, who are acquainted with your unwearied industry, in the advancement of philosophy, greedily expect the communication of your further experiments."[30] Harvey dismisses Ent's comment not because he did not believe the work would be of interest but because he had no desire to expose himself to public criticism. Why, Harvey asks, should he surrender the peaceful environs of his study only to "again commit myself to the unfaithful ocean?" This metaphor of the sea echoes the concerns of William Gilbert, when he asked why he should bother add to "so vast an Ocean of Books by which the minds of studious men are troubled and fatigued." But for Harvey, this ocean is not merely vast, it is unpredictable. To Ent he continues, "You are not ignorant, how great troubles my lucubrations, formerly published, have raised. Better it is, certainly, at some times, to endeavor to grow wise at home in private; then by the hasty divulgation of such things . . . to stir up tempests, that may deprive you of your leisure and quiet for the future."[31] Despite Harvey's pleas for solitude, Ent prevailed upon him to share his work. It was then that he discovered Harvey had a complete treatise on generation, "a work framed, and polished with very great pains," but gathering dust. Ent urged him to publish it, arguing both for the utility of the work to the anatomical world and the importance of it in affirming Harvey's honor and reputation. He also guaranteed that Harvey would not "sustain any further trouble in the business; for whatever of care is requisite to the oversight of the press, I shall willingly take wholly upon myself." Harvey reluctantly agreed. He had hoped to augment the work with an additional section on the generation of insects, but those observations had been destroyed by the Parliamentary army when it ransacked his London home. Despite his misgivings, Harvey relinquished control of the manuscript and told Ent that could either publish or suppress the work as he pleased.

When Ent returned home to study the work, he found it highly developed and beautifully argued. That it had been nurtured for so long out of the public eye was, to Ent, a tragedy averted only by his intercession. He closes his preface with a nod to his own editorial role in the publication: "But, of my self, I shall add only this much; that in this great business I have performed no more than the meer office of a midwife: producing into the light this noble issue of his brain, in all its parts and lineaments perfect and consummate, as it is now presented to your view, but staying long in the birth and fearing perhaps, some injurious blast of envy or detraction." He continues, saying that he elected to "oversee and correct the press . . . against the errors of the compositor," work that mirrored the efforts of correctors in printing houses. In Harvey's case, his notoriously poor handwriting made this role even more critical. Ent acknowledges that Harvey's messy script, "which no one without practice can easily read," had stumped compositors in the past. It no doubt contributed to the numerous errors that appeared in the first edition of *De Motu Cordis*.

In preparing *De Generatione Animalium* for the press, Ent chose the well-known London publisher and bookseller Octavian Pulleyn and the printer William Dugard. Dugard was an interesting choice, and some historians suggest Ent purposely sought him out. He was a known Royalist printer, with sympathies toward the Crown that Harvey himself shared. The Interregnum government of the 1650s, however, had little tolerance for such opinions. Under pressure from Cromwell's government, which sought to eradicate opposition, Dugard was arrested in early 1650 and charged with printing seditious material.[32] His presses and printing apparatus were seized until, under pressure from friends such as John Milton, he issued a statement asserting his support for Cromwell's Protectorate. With his presses returned Dugard opened for business again. One of the first books he printed was Harvey's work on generation in 1651. But Ent's choice of Dugard was more than a reflection of shared sensibilities, linking an author and printer who supported the monarchy. It was an attempt by Ent to insulate the content of Harvey's book, which Ent knew would be controversial. Deep in the treatise, after hundreds of pages devoted to the physiology of reproduction, Harvey's narrative takes a detour.

Chapter 68 begins with a discussion of comparative anatomy, explaining the uterine structures of rabbits, dogs, and other animals. Then the ground shifts:

And whilst I speak of these matters, let gentle minds forgive me, if, recalling the irreparable injuries I have suffered, I here give vent to a sigh. This is the cause of my sorrow:—whilst in attendance on his majesty the king during our late troubles and more than civil wars, not only with the permission but by command of the Parliament, certain rapacious hands stripped not only my house of all its furniture, but what is subject of far greater regret with me, my enemies abstracted from my museum the fruits of many years toil. Whence it has come to pass that many observations, particularly on the generation of insects, have perished, with detriment, I venture to say, to the republic of letters.[33]

This reminiscence is immediately followed by the header for chapter 69, on the uterine changes in a doe in the first month of pregnancy. Such a potent but hidden lament is not accidental. Buried in his work, Harvey lets loose his enduring frustration at the harm done to his scholarship by the Parliamentary army. To Ent, who was sensitive to the politics of the Interregnum, the passage signaled the need for caution. He therefore sought out a printer who would be willing to produce such a work, someone sympathetic to the criticism nested in its pages.[34]

When *De Generatione Animalium* came off the press in 1651, its title was presented in an allegorical frontispiece. Jove, replete with gilded horns, sits atop a pillar. His left hand cups the bottom half of an egg, while the right hand lifts off the top half. The egg is inscribed with the phrase "Ex ovo omnia" (everything from an egg), and out of it emerges a deer, an alligator, a grasshopper, and a human, among other creatures. To Jove's right sits an eagle, commonly considered his messenger and the creature that carried out his wishes.

Engraver Richard Gaywood is likely responsible for producing the image, though it is unclear who designed it. Harvey had stepped away from the project, so there is reason to believe Ent had a hand in the frontispiece. We could well read the image as not only an homage to Harvey, whose work on fertility ties into Jupiter's role, but also to Ent as his assistant, the eagle carrying out the will of his colleague and linking Harvey to the rest of the world of learning. Ent's work on Harvey's behalf—which took place over at least three years—involved convincing his colleague to publish, securing a printer and publisher, and then preparing the manuscript copy for the press. His efforts were not

William Harvey, *De Generatione Animalium* (1651).
Courtesy of the Hunting Library.

those of a traditional Renaissance corrector, nor was he acting as a publisher, since there is no indication that he financially subsidized the work. He was, instead, operating in a new role that responded both to authorial reluctance toward print and to the demands of seeing a work into print. There was no title for the job Ent assumed when he visited Harvey that day in 1647. As we shall see, however, it was a job that would be repeated by others in the scientific world.

Van Schooten and the Geometers

If Ent demonstrated a new range of editorial skills in preparing the work of a single individual for the press, the next example illustrates how an editor's influence could be felt across an entire field. Dutch mathematician Frans van Schooten the younger (1615–1660) was arguably the most influential editor of early modern mathematics, and his work affected a generation of productive mathematicians, both in the Low Countries and beyond. He was a prodigious mathematics student who studied under his father, Frans senior, a professor at Leiden University and the Leiden Engineering School.[35] In 1637 he met René Descartes and learned of the Frenchman's pending mathematical treatise, *La geometrie*. The work was in the process of being published in Leiden as part of Descartes' larger *Discourse on Method*, and van Schooten studied the proofs as they came off the press. He was even employed by Descartes to do some of the mathematical diagrams in that first edition.[36] And although van Schooten struggled with Descartes' dense analytical demonstrations, he quickly recognized the power of the work. Over a period of years he set out to edit and translate it. Meanwhile, with a letter of introduction from Descartes and financial support from the Elzevier publishing house, van Schooten visited Paris. There he met Marin Mersenne and was shown unpublished manuscripts by Pierre de Fermat and François Viète, two of France's leading mathematical minds. Van Schooten realized immediately the value of the works. He copied them by hand, returned to Leiden, and proceeded to edit Viète's work for the Elzeviers, who published his *Opera Mathematica* in 1646.

Riding the momentum of this accomplishment, van Schooten turned his efforts back to Descartes' *La geometrie*. Originally written in French, he knew the work needed to be in Latin to have any meaningful impact on the mathematical community. Van Schooten completed this translation in 1649, and his efforts helped establish a tradition of Cartesian mathematics in the Low Countries. More than putting the work into the scholarly language of the day, van Schooten transformed Descartes' dense, often vague treatise into a clear textbook, replete with geometric diagrams that he produced himself.[37] There were also entire sections that went through the proofs, explaining steps that Descartes had glossed over in his original treatise and linking the theoretical ideas to concepts in other texts. He also generalized Descartes' ideas, so that readers could see how the geometric theory could be more broadly applied. For example, in discussing aspects of conic sections (ellipses, parabolas, hyperbolas), van Schooten inserted his own italicized comments, addressing the reader directly. "And since this way of finding these lines [i.e., the normals to the three curves discussed by Descartes] is not only elegant and subtle, but also interesting in its own right, I trust that it will be welcome to those who want to exercise with these things, if I show how they are to be found for the Hyperbola and Parabola, and also for the conchoid."[38] After the first edition of his translation was praised by mathematicians, van Schooten expanded the scope of the second edition by using the work of his students and peers to clarify the application of Cartesian methods to classical problems. The contributors included Florimond de Beaune, Johan de Witt, Jan Hudde, Hendrik van Heuraet, and the young and promising mathematician Christiaan Huygens.[39] The editing of Descartes' work reflected van Schooten's mathematical and pedagogical skills, and it facilitated the widespread adoption of Descartes' methods. This was not the traditional work of a corrector. Van Schooten was editing in a more modern sense.

Van Schooten's 1649 translation was published in Leiden by Jan Maire, as Descartes' original had been. A second edition appeared in 1659. Both the 1649 and 1659 editions of *Geometria* have separate title pages that introduce the various mathematicians who contributed demonstrations to the work. In neither edition is Huygens listed, but on pages 203–5 in the 1649 edition van Schooten writes, "Another example, which I put forth here for consideration, I selected from the invention of the noble and distinguished young man Christiaan Huygens."[40] Despite not having a marquee position in the book's index, Huygens had come to the attention of both Descartes and Mersenne, who recognized his prodigious skill in mathematics and referred to him as a "young Archimedes." Inclusion in van Schooten's work put Huygens on the map. Subsequent editions of the Latin *Geometry* were published by the Elzeviers between 1683 and 1695, again with supplementary commentary by van Schooten and work by students added for explanatory purposes. As each edition built on the previous one, the work grew from Descartes' original 116 pages to more than 500.[41] It was this pedagogical influence on the shape of the work that made it the standard geometry textbook of the period.

Van Schooten's efforts on behalf of Descartes were further amplified in his role as editor for Christiaan Huygens' mathematical treatises. Huygens was still

a relatively young scholar when he began sending his work to his former in-
structor, asking for criticism. When Huygens completed a treatise he likewise
sent the draft to van Schooten, who read with a critical eye, correcting errors,
pointing out redundancies, and generally offering suggestions for how arguments should be structured.[42] He played an essential role in seeing Huygens'
first mathematical treatise, *Theoremata de Quadratura* (1651), into print. In September of that year, Huygens sent a draft of the work to van Schooten, who
returned it with corrections, structural comments, and the following note:

> I wish that you would publish it as soon as possible, to show everybody that you
> [are] the first with the hyperbola . . . You can publish the treatise as an intro-
> duction to your other work and because it is short, have it printed in a rather
> large letter and in a small size, octavo for example; especially since little has to
> be changed and the figures are nicely drawn, it is very suitable for such a format.
> While reading I found very few changes necessary, except that I omitted, added,
> or changed a word here or there so that the sentence became clearer to me.[43]

Huygens made the suggested changes and two months later wrote his editor again: "The first part you read recently, and for that reading I will forever
be indebted . . . The other part I now send to you in polished form that, if you
please, you can read it over, primarily considering my arguments and whether
you believe them skillful enough to be thought intelligent."[44] Two days later van
Schooten had read and responded to the draft, telling Huygens that—although
it demanded one's attention and required effort to understand—he found the
demonstrations to be very pleasing.[45]

Huygens' second mathematical treatise, *De Circuli Magnitudine Inventa*,
dealt with the properties of the circle. This, too, he sent to van Schooten for
editing. In April 1654 Huygens wrote, "In my treatise, which I recently gave
you to read, I noted one demonstration to be lacking a rule which is in the
final problem. I now send that demonstration . . . Truly this little book at some
point is going to be extensive, and will be abundantly supplied if I add the
construction of those illustrious problems that seem pleasing to you. Beyond
these, great van Schooten, I submit to your counsel. If you disagree with anything, by your great judgment, it will be changed."[46] Just over a week later, van
Schooten wrote back with praise for the work and encouraged Huygens to
include two other problems. He also facilitated the publishing process for
Huygens by getting in touch with the printer: "Yesterday I highly recommended
the book to Daniel Elzevier himself . . . I told him that you had approached

other printers in The Hague, who were also interested in printing the work
. . . but I told him you preferred the Elzevier printers . . . I also emphasized
that you wanted no expensive figures, and that you expected fifty copies for
the original manuscript, which truly, in all confidence, they would not refuse
at all."[47] Elzevier assured van Schooten that he would return to The Hague the
following week to speak with Huygens directly about the project, but clearly
the arrangements were made for Huygens ahead of time. The book was published and copies bound in June 1654, and as a final request, Huygens asked
that van Schooten distribute copies to certain mathematicians in the Low
Countries. With these two publications, Huygens established himself as one
of Europe's preeminent mathematicians. Nevertheless, we need to recognize
van Schooten, who was in the background reviewing the manuscripts and providing feedback, as a catalyst to Huygens' publications and reputation. His invisible hand guided the works from the desk of a young mathematician to the
renowned press of the Elzeviers.

Van Schooten's role in editing Huygens' publications continued with the
treatise *On Probabilities in Games of Chance*, an important work in the history
of probability. Huygens was inspired to write it after learning of about a particular problem—related to odds in a dice roll—that he had encountered on
a visit to Paris in 1655.[48] Huygens, Blaise Pascal, and Pierre de Fermat continued working on the problem through correspondence, each mathematician
calculating expectations and sharing related queries.[49] In the spring of 1656
Huygens wrote to French mathematician G. P. Roberval, "I have since . . . written the foundations of the calculus of the games of chance at the request of
Mister Schooten who wishes to print it."[50] Two days later he wrote van Schooten: "Here you have the work on games of chance that you requested, though
written in the vernacular, which was necessary for me to do, the Latin terms
failing to work."[51] By mutual agreement, the treatise would not appear as a
stand-alone publication—in part because of its brevity—but would be appended to a work van Schooten himself was about to publish, entitled *Mathematical Exercises*. Huygens' treatise appears in the fifth section, "Sectiones
Miscellaneae Triginta," having been translated by van Schooten into Latin.[52]
It does not have its own title page, but Huygens' name appears in the table of
contents, and his fourteen-page "De Ratiociniis in Aleae Ludo" ("On Probabilities in Games of Chance") has its own preface to the reader, written by van
Schooten. This preface is followed by a dedication Huygens penned, where
he describes the history of the joint publication and recognizes van Schooten's

efforts in translating and editing the work for publication. Once printing was concluded, van Schooten took the lead in circulating the treatise. He sent a copy to the English mathematician John Wallis, as well as to the French contingent of Pierre de Carcavy, Claude Mylon, and Pierre de Fermat. Huygens may have received a few personal copies to circulate, but most mathematicians who commented on the book seem to have received their copy directly from the primary author, van Schooten. A second edition of the work, this one in Dutch, was published in 1660 by Gerrit van Goedesberg of Amsterdam, and again van Schooten was responsible for the translation. His numerous efforts on Huygens' behalf, on this and the earlier publications, speaks foremost to the level of commitment he had to his former pupil, but it also reminds us how essential his role was to Huygens' success. The extent to which he was involved in seeing Huygens' work into print, even after he had an established reputation as a premier mathematician, is hard to overstate.

Robert Sharrock and Robert Boyle

In early 1660 Henry Oldenburg wrote a wide-ranging letter to Robert Boyle, updating him on recent scholarly events. He remarks on advances made by German chymists, and new anatomical studies coming out of France. He also mentions "my desire to see the treatises on plants and on morals."[53] Oldenburg had referenced these works before in a letter to a different colleague but never with attribution. A month later, Oldenburg's curiosity was apparently satisfied by Boyle, because he wrote to thank him: "I am much obliged to you for your liberality of presenting me with those two desired treatises of plants and manners." Both works, it turns out, were written by Robert Sharrock (1630–1684), who was beautifully summed up by one contemporary as "learned in divinity . . . and knowing in vegetables."[54] Sharrock was a clergy member and naturalist who took his doctorate in law at Oxford in 1661.[55] In addition to his legal and theological work, he devoted significant time to botany and developed a strong interest in the experimental approach to philosophy. The Physic Garden at Oxford provided him with a living laboratory, allowing him to focus on seeds, seedlings, and propagation, and his output has been likened to that of famous botanist Nehemiah Grew.[56] Sharrock knew the scholars who comprised the Oxford Philosophical Club, an informal but intellectually active group that included Boyle, John Wilkins, John Evelyn, and Christopher Wren. Wilkins invited Sharrock to join them in 1653. These men met at each other's homes, always with the hope that, as Sharrock wrote, "we shall have some endowment for a society to study experimental philosophy at Oxford, which I am very glad of."[57] But even prior to the group's meetings, Sharrock and Boyle had developed a close working relationship from which both men benefitted.

In 1660—as Oldenburg's letter noted—Sharrock published *De Officiis secundum Naturae Jus* (*Duties according to the Laws of Nature*) and *The History of the Propagation and Improvement of Vegetables*. Both works were dedicated to Boyle, who certainly encouraged Sharrock intellectually and may have supported his publications financially as well. The former treatise dealt with morality and was a critique of Thomas Hobbes' ethics, while the treatise on propagation presented important ideas in the history of botany.[58] Printed by Ann Lichfield for the Oxford bookseller Thomas Robinson, it was popular enough among horticulturalists to go through three subsequent editions.[59] In Sharrock's dedication to Boyle, he thanks his friend while hitting an appropriate note of self-deprecation: "I imagined no more being an author, or compiler of any matter on this subject, than of doing any other thing which I have neither fancy nor fitness to do."[60] It was Boyle's encouragement, Sharrock says, that helped him bring the treatise to fruition. His gratitude was more than rhetorical, however, since Sharrock had the unique opportunity to reciprocate and assist Boyle soon after he published his own works. After years of collecting papers and compiling manuscripts, Boyle was—by 1660—ready for an extensive publishing run. Between 1660 and 1666 he published eleven treatises in multiple editions, both English and Latin. Boyle understood that the press was a critical instrument in the dissemination of knowledge, but his attitude toward print can only be characterized as tepid.[61] Part of the circle of scholars connected to Samuel Hartlib, Boyle was regularly engaged in the exchange of letters and manuscripts. As we saw in chapter 2, many of the individual essays that made up Boyle's *Certain Physiological Essays* had been composed and shared with his colleagues at least six years before they were published.[62] His eventual acquiescence to publish was related to the Restoration.

When Charles I was executed in Whitehall, London, on January 30, 1649, it was the culmination of years of bloody civil war between the Crown and Parliament. Until 1660, England operated under a lord protector, rather than a king. Initially this was Oliver Cromwell, the erstwhile leader of the Parliamentary army, and only upon his death did his son succeed him. Finally, in 1660, Charles II returned from exile and was restored to the throne, whereupon he immediately ramped up censorship of printers with the 1662 Licensing of the Press Act.[63] This led to a dramatic drop in the number of published pamphlets and books, but it added legitimacy to those works that saw their way to print. Boyle exemplified the kind of individual—a gentleman and a scholar—whose work could carry the government's imprimatur. In the same year the Licensing Act was passed, the king granted a charter to the Royal Society, a body dedicated to the exploration of natural philosophy. This legitimization helps explain Boyle's willingness to see his work into print after 1660. But even as he did so, he found he needed more than an amanuensis to prepare his treatises for publication. He needed an editor, and the first person to assume the role was Robert Sharrock, though others—including Oldenburg—would eventually assist Boyle as well. I focus on Sharrock's efforts because they widen our understanding of early modern editors and they draw into sharp relief the degree to which scientific authors relied on them.

Sharrock's editorial contributions began with Boyle's *New Experiments Physico-Mechanical, Touching the Spring of the Air*. In January 1660 he wrote to Boyle about his work, "This day I consulted Mr. Robinson about the printing of your piece [*New Experiments*], the paper and character and time requisite for the work."[64] Sharrock, like Ent and van Schooten before him, worked directly with Robinson's printer, Henry Hall, on Boyle's behalf. At roughly the same time Robinson was also publishing Sharrock's botany treatise, so assisting Boyle with his work made practical sense. But it also speaks to the level of attention he gave his colleague's work, seeing to such details as the appearance of the text, the quality of paper, and the timing of production, which would result in five hundred copies of the English edition. Sharrock informed Boyle that they could expect two sheets to be printed weekly, and he promised to help with correcting printed pages. He had the time, he said, since he was finished with his own publication. Sharrock was also enlisted to translate some of Boyle's works into Latin. In his "Preface to the Friendly Reader" in *New Experiments*, Sharrock writes, "Because the noble author is willing to oblige all men, he has already provided, that this piece shall shortly be done into Latin, that so it may come home to diverse worthy persons in its stream, who cannot travel to find it out in its first origin."[65] Translation was underway when he wrote that, though for efficiency's sake Sharrock had delegated some of the work to assistants. A few months later he wrote Boyle, expressly unhappy with the translation work done by these assistants:

> I shall be willing to employ a very considerable part of my time upon any business you shall employ me. I have employed several hands in your translation

but . . . the whole matter being so different from that they are ordinarily conversant with (which is generally the vulgar philosophy, humanity & divinity) it happens that after much assistance given them as to words and phrases their help is so inconsiderable that in some [instances] I have been forced to alter every period in every sentence. However I hope & question not but this will be translated for you ready for the press by midsummer.[66]

Sharrock emphasized his commitment to the project, explaining that he would only attend to his studies half of the time; the remainder would be devoted to Boyle's treatise. "I shall do no other business but yours in this overlooking of copy, the press, and rendering them in Latin."

By November Sharrock reported that he had nearly completed the translation of *New Experiments*. He suggested that Boyle hire an amanuensis—someone specially trained to write from dictation or a manuscript copy—to transcribe the work for the printer so that the compositor had a clean copy to read. Sharrock was eager to see this happen and wrote again two weeks later, asking Boyle to "expedite the sending of copy" so that the work could be printed. This urgency persisted through the letter, Sharrock pressing the point that if Boyle did not promptly return the copy to him, it would put the treatise at risk for piracy abroad, making the printer unhappy.[67] Meanwhile, Sharrock continued his editing. He wrote to Boyle in mid-December 1660 about an editorial change: "I have taken a little liberty in the translation of your plaister of lime and cheese, but so as to make it more intelligible, having often made it good and as often bad till I had learn'd the knack of it."[68] Not only was Sharrock translating Boyle's work into Latin; he brought to the task his own experience in chemistry. This indicates a much more robust role than corrector or even translator. Sharrock's work was rooted in knowledge of the discipline. This familiarity, however, made him less forgiving when others fell short. He took issue with the translation work being done for Boyle by Henry Oldenburg, who stepped in to help when Sharrock had to be away. His concerns seem minor but likely reflect his sense of ownership where Boyle's work was concerned. For example, he disliked the fact that Oldenburg translated the word "sucker" (part of an air pump) differently in multiple places, sometimes using the Latin *embolus* and other times *sipho*. Sharrock suggested Boyle address the inconsistency in the errata or perhaps in a general note, but either way, he wrote, "both will be blemishes upon the sale of your book and no credit to the translation, wherein for your honours sake I must not be unconcerned."[69] Another solution he offered involved reprinting the offending page and a half, and paying for the cost out of pocket. He explains, "Were it the printer's fault I would make him reprint it at his own charge, but his print is well enough according to the copy. The copy is as it was sent to Mr. Robinson for the press. The only mistake was in Mr. Oldenburg who sent down the copy." This critique of Oldenburg's work signals the degree to which Sharrock identified himself as Boyle's primary editor, one who was not interested in being displaced. His Latin translation of Boyle's treatise, *Nova Experimenta Physicomechanica*, was published in 1661.

That year also saw Boyle's publication of two other works: *Certain physiological essays* and *The Sceptical Chymist*. Sharrock had a hand in both works, and in both he referred to himself as the publisher. At the time, being a publisher often implied a fiscal obligation, but we have no evidence that Sharrock provided any monetary support for Boyle's work. Indeed, this would have been unlikely and inappropriate given Boyle's family wealth and standing. Sharrock adopted the term out of necessity; nothing else quite captured his efforts to shepherd the treatises into print. In his preface to *Certain physiological essays*, he outfits the reader with a framework for understanding Boyle's essays, which had been composed in the mid-1650s, circulated in manuscript form, and then brought to the press. For the *Sceptical Chymist* he did not include a preface, only a brief statement at the very end of the treatise promising readers that he, "the publisher," would have a Latin edition of the work available soon.[70] That translation appeared in 1662, the same year Boyle also published *A Defence Of the Doctrine touching the Spring and Weight Of the Air*. Sharrock is once again present in the preface—"The Publisher to the Reader"— where he accounts for the delay in this publication, alludes to the printer's eagerness to see it finished, and then apologizes for the errata in the work. Boyle, he says, is not one for correcting and revising because of his weak eyes and, "neither could the publisher [Sharrock] much attend the press, it being printed in a distant place from his usual abode."[71]

Sharrock's longest preface came in Boyle's 1663 *Some considerations touching the usefulnesse of experimental natural philosophy*. Here, he advocates for Boyle's place in the broader framework of Christian ideas in the seventeenth century. With characteristic rhetorical skill, he suggests that Boyle's treatise would almost single-handedly bring honor to the age and that his work in natural philosophy could bring readers closer to God and augment their ability to do good deeds here on earth. "As soon as I had the author's leave," Sharrock wrote, "I durst not forbear the committing of them to the press." He bemoans Boyle's habit of losing papers but praises print for capturing and making permanent ideas that would otherwise be lost in the shuffle. Sharrock viewed his role as that of preserver, and his instrument was the press. For Boyle, Sharrock was much more than that; he was a combination of archivist, curator, and editor, seizing upon Boyle's treatises as soon as he could and working tirelessly to see them into print.

Edmund Halley and Isaac Newton

Through Ent and Sharrock we begin to see the editorial role take shape, but arguably no individual exemplified the characteristics and skills required better than Edmund Halley, whose work on behalf of Isaac Newton facilitated the publication of one of science's most important works. As Rob Iliffe has shown, anonymity and solitude formed the basis of Newton's publishing strategy, and the evidence bears this out, as Newton pleaded time and again for his name to be removed from theories and proofs.[72] It was more than just the spotlight he sought to avoid but the inevitable debates that accompanied it. It was this posture that gave rise to one of the most famous examples of "publishing midwifery" in the early modern sciences. Halley's work on behalf of Isaac Newton ensured that the *Philosophiæ Naturalis Principia Mathematica* (*Mathematical Principles of Natural Philosophy*) made its way to press, thereby introducing to the world Newton's theory of universal gravitation. Acting in much the same vein as other editors we have discussed, Halley assumed a pivotal role in turning Newton's magnum opus from a manuscript into one of the most important publications in the history of science.

Newton's reliance on others to see his work into print has hardly gone without notice. Without Henry Oldenburg, for example, none of Newton's work leading up to the *Principia* would have been published. Oldenburg saw to the printing of numerous treatises, especially on the subject of optics. But Oldenburg did more than simply include Newton's letters and demonstrations in the Royal Society's *Philosophical Transactions*. He also used his reputation to confer credit on Newton's work.[73] He was, in effect, an arbiter of quality scientific contributions, and he brought the Cambridge mathematician into a conversation with top scholars. Once there, Newton found much of the intellectual exchange fruitful, but it did little to quell his desire for solitude. Several heated exchanges over priority—in particular with Robert

Hooke—only affirmed his desire to keep out of the limelight. It is not surprising, then, that it took some effort for Halley to coax Newton's greatest work from him.

The story of Halley's initial visit to Newton, familiar to most historians, has all the elements of great theater. The genius professor, cooped up in his Cambridge house amid stacks of papers and books, is visited by a London astronomer who comes in the name of a mathematical query. Specifically, Halley made the trip to ask Newton whether he knew what path the planets would follow—or what kind of curve would be described—if they were attracted by the sun with a force that varied inversely to the square of the distance.[74] Newton did not hesitate: the path would be an ellipse. He claimed to have calculated it in a paper some five years earlier, but, alas, he could not find it at that particular moment. Newton promised Halley he would rewrite the proof and send it to London. He did so a few months later, sending a short treatise entitled *De Motu (On Motion)*, Newton's first public explanation of celestial mechanics.[75] Halley recognized the value of the work and shared it with the Royal Society, whose members were enthusiastic about its promise. Bolstered by the response from the society and some esteemed mathematicians, Halley paid a second visit to Cambridge, this time urging Newton to expand on the treatise and produce a more complete treatment of planetary motion. The trip was reported in a Royal Society meeting of December 1684: "Mr. Halley . . . had lately seen Mr. Newton at Cambridge, who had shewed him a curious treatise, *De Motu*; which, upon Mr. Halley's desire, was, he said, promised to be sent to the Society to be entered upon their Register. Mr. Halley was desired to put Mr. Newton in mind of his promise for the securing [of] his invention to himself till such time as he could be at leisure to publish it."[76]

This visit marks the moment that Halley shifted into the kind of editorial role I have explored in this chapter, as he shepherded Newton's work into print by actively assisting in numerous ways. He edited drafts and read proofs for clarity and accuracy. He dealt directly with the printers and engravers, and he tended to the many aesthetic decisions that needed to be made. Finally, he supported the endeavor financially, at great personal risk.

De Motu was a critical first step in the composition of Newton's larger work, but the heavy lifting was yet to come.[77] Of three extant copies of *De Motu*, one is in Halley's hand, indicating that at some point he transcribed the work for Newton. But as Bernard Cohen has demonstrated, the version Halley copied was incomplete, lacking two problems related to motion in resisting media that are found in the final version. Halley addresses these missing problems in a letter to mathematician John Wallis in December 1686: "Mr. Isaac Newton about 2 years since gave me the inclosed propositions, touching the opposition of the medium to a direct impressed motion . . . I begg your opinion thereupon, if it might not be (especially the 7th problem) somewhat better illustrated."[78] There is also a partial manuscript draft of the *Principia* at the University of Cambridge that contains Halley's page-by-page, line-by-line changes and comments. They appear on the recto, while Newton's responses to the editorial comments appear on the verso. Suggestions include phrasing for specific lemmas, propositions, or definitions, while others offer alternative ways of presenting the mathematics. We know from subsequent drafts—thanks to Bernard Cohen, who fastidiously put them in order—that some of Halley's proposed changes were incorporated, while Newton passed over others. This was a detailed, mathematically rigorous editorial effort on Halley's part.[79]

Halley's work, however, extended beyond the mathematical details. On April 28, 1686, Newton's draft of the *Principia* was presented to the Royal Society, who enthusiastically called for its publication. In Halley's subsequent letter to Newton, he writes that the society "resolved to print it at their own charge, in a large quarto, of a fair letter," but the issue of who would pay for the printing was hardly clear cut. Key members of the society, including the president, were absent from the meeting so it was impossible to determine who would subsidize the work and at what rate. To keep the process moving, printing was handed off to Halley: "I am entrusted to look after the printing it, and will take care that it shall be performed as well as possible, only I would first have your directions in what you shall think necessary for the embellishing thereof, and particularly whether you think it not better, that the schemes should be enlarged, which is the opinion of some here."[80]

Less than a month later, Halley sent Newton the first proofs, explaining that the paper and font Newton saw there were what they hoped to use for the final publication, unless Newton had some objection. He also assured Newton that while the printed type was not as clear as it should be, that was due to the newness of the font. "I have seen a book of a very fair character, which was the last thing printed from this set of letter; so that I hope the edition may in that particular be to your satisfaction." Once Newton expressed satisfaction with Halley's design choices, approving of the type and the paper, Halley wrote that he would "push on the edition vigorously." He also discussed the illustrations that would be incorporated with the text. Printers handled mathematical images in different ways, some including fold-out pages containing the geometric representations, others integrating them into the text. Halley hoped for the latter: "I have sometimes had thoughts of having the cuts neatly done in wood, so as to stand in the page, with the demonstrations, it will be more convenient and not much more charge, if it please you to have it so, I will try how well it can be done."[81] Newton's response included two geometric diagrams that he wanted Halley to incorporate if the images were going to be inline with the text. In the same letter discussing fonts, Halley urges Newton to complete the third part of the work, "The System of the World," which he says, "is what will render it acceptable to all naturalists, as well as mathematicians; and much advance the sale of ye book."[82] Sales were important to Halley, a consideration I discuss below. He hoped to see the book tie together Newton's complex mathematics by applying it to the structure of the universe. In closing he asks Newton to check the proof carefully in case he had missed any errors.

One book at a time Newton sent drafts to Halley, and one book at a time Halley returned the proofs to him for final approval. In February the project was bogged down—a printer had lost parts of a proof—but Halley writes Newton, "I will now do nothing else till the whole be finished . . . and to redeem the time I have lost, I will employ another press to go on with the second part, which I am glad to understand you have perfected; and if you please to send it up to me, as soon as I have it I will set the printer to work on it."[83] Halley then shepherded the work through two presses, obtaining proofs from each, connecting them, and fixing myriad errors caused "by the negligence of the printer." He was still, however, commenting on substantive points. In a letter of April 1687, regarding Newton's theory of comets, he writes, "I do not find that you have touched that notable appearance of comet's tails . . . but a proposition or two concerning these will add much to the beauty and perfection of your theory of comets." The flattery embedded in Halley's letter points to yet another aspect of his editorial role: that of moral support. Newton famously avoided publication and continually threatened to keep his mathematical ideas to himself if publishing led to conflict or inconvenience. Thus, a priority dispute, kicked off by Robert Hooke shortly after *De Motu* reached the Royal Society, was no small matter for Halley. Between reading proofs and dealing with printers, he also had to assure Newton that Hooke's claim—that *he* had discovered the inverse-square force first—was both untrue and ignored by the rest of the scientific community. Entire letters between Halley and Newton are devoted to allaying Newton's concerns.

> I am heartily sorry that in this matter, wherein all mankind ought to acknowledge their obligations to you, you should meet with any thing that should give you disquiet, or that any disgust should make you think of desisting in your pretensions to a Lady, whose favors you have so much reason to boast of. Tis not she but your rivals . . . that endeavour to disturb your quiet enjoyment . . . I hope you will see cause to alter your former resolution of suppressing the third book, there being nothing which you can have compiled therein, which the learned world will not be concerned to have concealed.[84]

Finally, Halley's involvement in Newton's publication led him to take on an additional, and more precarious, role, that of underwriter. On June 2, 1686, the Royal Society recorded the mandate: "It was ordered, that Mr. Newton's book be printed, and that Mr. Halley undertake the business of looking after it, and printing it at his own charge; which he engaged to do."[85] The society balked at paying for the work itself. It had previously financed the publication of Francis Willoughby's *History of Fishes*, only to find the market for such work was decidedly limited. Printing for the work had cost the society £406, but it recovered only £111 in sales.[86] Still smarting from that deal, the society's council was reluctant to enter another agreement for publication that might result in a similar deficit. The burden was passed to Halley, who, in 1686, was struggling financially himself.[87] At one point, Newton wanted to remove book 3 from the *Principia*, leaving only the first two mathematical books in the work. As editor, Halley's arguments against this change prevailed. Newton subsequently admitted that maintaining the book's accessibility would help sales of the work, which Halley anticipated with the hope of recouping his investment.

In the end, Halley's editorial influence was significant for its timing as much as his correction or work with the printers. Some historians have rightfully pointed to his light hand where Newton's proofs are concerned. There is, in fact, little trace of Halley's influence on the philosophical and scientific aspects of the work, but this does not diminish the importance of his role. In Halley's "Ode to Newton," which appears at the beginning of the *Principia*, the editor expresses his dedication to both his colleague and the treatise they brought forth. Newton, for his part, acknowledges his debt to Halley in the prefatory remarks: "In the publication of this work, Edmund Halley, a man of greatest intelligence and of universal learning, was of tremendous assistance; not only did he correct typographical errors and see to the making of the woodcuts, but it was he who started me off on the road to this publication."[88]

Halley imagined his reach would extend even further by controlling the distribution of the *Principia*, but the contract with the printer prohibited such a scenario and reined in what were no doubt financial ambitions of the astronomer.[89] His influence was nevertheless significant, and Newton could not have succeeded without him. Even after the *Principia* was published and Halley's work done, Newton continued to require assistance to carry his work to the press. In 1695, for example, John Wallis writes to Newton about his mathematical work related to calculus, "I have taken pains to transcribe a fair copy of your two letters, which I wish were printed."[90] Wallis says he was sending that copy to Newton for review, and he recommends having it printed at Oxford, where the printers had the required type and were familiar with such material. "And Mr. Caswell or I will see to the correcting of the press," Wallis assures him.

Historian of science A. R. Hall points to no fewer than ten individuals "who with or without Newton's own active cooperation saw his writings into print."[91] Given what we know about Oldenburg's efforts to publish Newton's early work, Halley's attention to the *Principia*, and the combined editing of men like Samuel Clarke and Roger Cotes later on in Newton's career, it is safe to assume that in the absence of an editor—someone to persuade Newton to print, to deal directly with printers, and to examine and correct proofs—the net output of Isaac Newton would have been significantly reduced. Even posthumously, editors and translators were critical to the Newtonian enterprise. John Colson, Lucasian Professor of Mathematics at Cambridge, translated Newton's *Method of Fluxions* into English in 1736. Until then, the work had remained in Latin, unpublished. Not only did Colson's translation make the treatise more accessible; his added commentary helped elucidate difficult passages for readers. In the preface, he describes his role as "only the Interpreter" but one who relished the pleasure of presenting to the public a work "of an elementary nature, preparatory and introductory to [Newton's] other and most arduous and sublime speculations, and intended by himself to for the instruction of novices and learners."[92] Colson's commentary on the work and his addition of annotations, illustrations, and supplementary examples rendered Newton's ideas accessible to an eager audience of mathematicians. A review in the *Philosophical Transactions* by John Eames praises Colson's translation. "And farther to explain this work, and to supply such things, for the use of common readers, which the author, according to his usual brevity, has often omitted; the translator has thought fit to give us a comment on a good part of the work . . . His fitness for such an undertaking is well known to the learned world."[93] The review continues for eight more pages, offering specific examples of how Colson augments Newton's original work. For example, Newton offers a general solution to find the root (or fluent, as Newton writes) of any proposed "Fluxional Equation." In modern parlance this is Newton's description of finding the solution to a derivative function, a foundational operation in calculus. Reviewing this section of the treatise Eames writes, "Mr. Colson in his comment upon this part of the work is very full and explicit. He explains and applies the author's particular solution; but is much more copious in explaining the examples, and clearing up the difficulties and anomalies of the general solution . . . The commentator concludes by giving us a very general method for resolving all equations, whether algebraical or fluxional."[94]

Colson's work translating and editing Newton was just one example of his efforts to make important authors and treatises more widely available and accessible. His entire career, in fact, was inclined toward pedagogy. Prior to his appointment at Cambridge he was master of the Free Mathematical School at Rochester, founded in 1701, and many of his publications were intended to guide young mathematicians in their study of analysis. These included *The Construction of Use of Sphaerical Maps* (1736), *Dr. Saunderson's Palpable Arithmetic Decypher'd* (1740), and Colson's translation of Abbé Nollet's *Lectures in Experimental Philosophy* (1752). His final and perhaps most infamous project was a translation of Maria Gaetana Agnesi's *Analytical Institutions for the Use of Italian Youth* (1748), which Colson took on late in life and for which he learned Italian to carry out the translation. Unfortunately, he wrongly credited Agnesi with discovery of a cubic curve, but the work nevertheless reached an English-speaking audience in 1801, some forty years after his death.[95] Colson was impressed with Agnesi's work and believed that young scholars in England should have access to it. According to the editor who ultimately oversaw publication, Colson had finished the project prior to his death "and had actually transcribed a fair copy of his translation for the press, and begun to draw up proposals for printing it by subscription."[96] But Colson had even more ambitious plans to edit the work specifically for a female audience. "In order to render it more easy and useful to the Ladies of this country . . . he had designed and begun a popular account of this work, under the title of *The Plan of the Lady's System of Analyticks*; explaining, article by article, what was contained in it. But this he did not live long enough to finish."[97] The draft can be found today in the Cambridge University Library, an unpublished manuscript containing sixty-one pages of Newtonian theory distilled for women.[98] Unfortunately, the lag time between its original publication and Colson's English translation doomed the book to a tepid reception. "It can do no good now, or, to speak more precisely, there are other books of a like nature and less bulk which can do *more* good," wrote the anonymous English reviewer in 1803.[99] Nevertheless, we see in Colson another example of the editor's influence on scientific publishing, this time bringing to a wider audience—of men and women alike—works that would otherwise have been difficult to parse, if not impenetrable.

Editing Science in the Age of Journals

Thus far we have considered those cases where an individual brought their editorial skills to bear on a colleague's project, ensuring that scientific ideas made it into print. But the seventeenth century also witnessed the rise of the periodical editor. As we saw in the previous chapter, Huygens' strategic use of journals reflected scientists' awareness that the publishing landscape was changing rapidly. With journals, scientists had a new way to use print to their advantage. The linchpin of a successful journal was its editor, someone who could carry out the double duty of acting as colleague and correspondent for their peers, while serving as a gatekeeper for these specialized conduits of information. The former was a well-established role. Marin Mersenne, Samuel Hartlib, and Henry Oldenburg cultivated extensive correspondence networks, and as postal systems became more robust, these networks operated with remarkable efficiency. More than one thousand letters would pass across Oldenburg's desk in a single year.[100] These letters were more than rudimentary starting points for the development of scientific or mathematical ideas; they played a critical role throughout the process.

We often assume a linear and increasingly formalized arc in the development of scientific ideas: scribbled notes become private letters that might become pamphlets on the way to being published as books. In reality scientific communication was much more complex and multidirectional. Letters from individuals often garnered epistolary responses from groups. Those responses might be compiled in a published book, which could later be addressed in additional letters or a printed broadside. The order of operations varied, and the emergence of scientific journals reflects this. They became the standard mode of publication in science because they were malleable, they were affordable, and they provided authors with a curated readership, typically members of scientific societies. We can situate these early journals in the liminal space between written correspondence and printed stand-alone works.[101]

When scientific journals first emerged, they were created from the very letters that were so readily exchanged among scientists. The Charter of the Royal Society laid the foundation for this in England. Fellows were granted the authority, "by letters or epistles . . . to enjoy mutual intelligence and knowledge with all and all manner of strangers and foreigners, whether private or collegiate, corporate or politic . . . in matters of things philosophical, mathematical, or mechanical."[102] Most of the entries in the *Philosophical Transactions* in its first five years were letters Oldenburg elected to publish there, and he had no clearly established standards for what was included.[103] In the first volume of the *Transactions*, for example, Oldenburg was the identifiable author of fifty articles while his colleagues accounted for only twelve.[104] This ratio shifted over time as the number of contributors increased, but it was still Oldenburg's task to solicit materials, cull relevant pieces for inclusion in the journal, and see to the technical details of the journal's printing. His editorial efforts overlapped significantly with those of Sharrock and his peers, and they deserve greater attention.[105]

The defining attribute of a good early modern editor was efficiency, a quality not infrequently linked to organizational skill. Henry Oldenburg's success had a great deal to do with the system he developed for managing a high volume of correspondence. Martin Lister, who frequently wrote the Royal Society about matters related to natural philosophy, recounts Oldenburg's process: "He told me he made one Letter answer another, and that to be always fresh, he never read a Letter before he had Pen, Ink and Paper ready to answer it forthwith; so that the multitude of his Letters cloy'd him not, or ever lay upon his hands."[106] The need for a system was clear. In the correspondence that passed through Oldenburg's hands during a single year, 1669 to 1670, there were at least forty manuscript works exchanged, many headed for journal publication. The contents of the 1670 *Philosophical Transactions* bear this out. There were twelve issues in this year (numbers 57–68), and communications to Oldenburg constituted thirty-five out of forty-five printed articles, or 78 percent of the journal's entries. Of the ten that were not direct letters to Oldenburg, two were translations by him of works printed elsewhere—such as in the *Journal des Sçavans* or the Italian *Giornale de Litterati di Roma*—or they were exchanges between two scientists whose conversation Oldenburg had mediated in some way. He had a hand in every entry that was printed. An illustrative example is his publication of Robert Boyle's experiments on respiration, which filled two complete issues of the journal (numbers 62–63). In introducing the work, Oldenburg writes, "These experiments . . . were by their Noble Author communicated to the publisher of these papers; who esteemed it more convenient to make them a part of these tracts (they taking up the room but of a few sheets) than to publish them any other way."[107] We have already looked at Boyle's collaboration with Robert Sharrock, but by the mid-1660s Sharrock had been replaced by Oldenburg. The latter directly facilitated the publication of such works as Boyle's 1664 *Experiments and Considerations Touching Colors*, which he also translated into Latin, and 1665's *New Experiments and Observations Touching Cold*. Oldenburg's editorial role was clearly established when he decided that the best forum for publishing Boyle's respiration studies was the *Philosophical Transactions*. The experiments that began in issue 62 continued in the next month's issue. This time, Oldenburg allowed for the inclusion of a preface (which he wrote himself), something not often seen in journal articles. He explains that the experiments described by Boyle were not initially written for the press, and thus, the reader should not expect polished and elegant language. Rather, they would encounter "novelty, and truth . . . impartially delivered." This is, of course, the kind of rhetorical posturing that one expects from scholars in the period. And the "matters of fact" that Boyle purports to present have already been considered in a broader sociological context.[108] The important element here is Oldenburg's role as both Boyle's and the journal's editor, and the way his personal relationship to the natural philosopher was integrated with—and enhanced by—his capacity as Royal Society secretary and publisher of the *Philosophical Transactions*.

In addition to the labor of selecting, organizing, and presenting scientific information, Oldenburg was fully responsible for financing the *Philosophical Transactions*, a position he clarified vehemently in the twelfth issue of the journal: "The writer thereof hath thought fit, expressly here to declare . . . that he, upon his private account (as a well-wisher to the advancement of useful knowledge, and a furtherer thereof by such communications, as he is capable to furnish by that philosophical correspondency, which he entertains, and hopes to enlarge) hath begun and continues both the composure and publication thereof."[109]

Oldenburg was keen to distance his efforts as journal editor from his work as secretary for the Royal Society. He acknowledged that he incorporated into the journal both papers and ideas that were presented to the society but clarified that the journal itself was independent of it. It was Oldenburg's job to ensure the printing of one thousand copies per issue, which would cost around fifteen pounds. But the monetary side of things caused him significant grief. After two years of editing the journal, he wrote to Boyle to express concern that people believed he profited greatly from its publication. "But I will make it out to any man, that I never received above 40£ a year upon [the journal's] account (and that is little more, than my house-rent)."[110] Oldenburg had published 136 issues of the journal when he died, whereupon Edmund Halley assumed the position. Like his predecessor he was economically responsible for the *Transactions*, a burden he tried to mitigate by ensuring that the society purchased the first sixty copies of each issue. It was only in 1752 that the society finally assumed full fiscal responsibility for the journal that was so well known and so inextricably linked to it.

In his study of early modern print shops, Adrian Johns acknowledges that terms like *bookseller, wholesaler, publisher*, and *editor* were relatively unfamiliar in the period and when used, tended to have idiosyncratic meanings.[111] Like Robert Darnton, whose model has shaped discussions of print culture for generations, Johns considered the participants in print culture on their own terms. In this chapter I have attempted to understand the kind of work that individuals carried out when shepherding a scientific work into print. It was a role that had many labels, none of which fully captured the extent of their efforts. They acted as more than correctors or proofreaders; their support was more than financial. Editors of scientific works motivated authors to commit their ideas to the paper, they steered that material into presses, they translated it into other languages, and they guided the works in postproduction, facilitating their reception among ideal readers. Halley, Ent, van Schooten, and others are hardly invisible in the history of science, but much of their work as editors is relegated to the periphery. They carved out the tasks that were only later codified as editorial work, and they played a consequential role in bringing scientific authors to print.

Reluctance Overcome

I n John Locke's *Essay concerning Human Understanding*, he pointedly identifies the roles of both the scientist and philosopher in society:

> The Commonwealth of learning, is not at this time without master-builders, whose mighty designs, in advancing the sciences, will leave lasting monuments to the admiration of posterity; but every one must not hope to be a Boyle or a Sydenham; and in an age that produces such masters, as the great Huygenius, and the incomparable Mr. Newton, with some other of that strain; 'tis ambition enough to be employed as an under-laborer in clearing the ground a little, and removing some of the rubbish, that lies in the way of knowledge.[1]

Locke understood the gap between the printed communications of natural philosophers—as recorded in papers, journals, and books—and the general educated public that was interested in those ideas. He made the above observation, however, as part of a broader discussion of language, ideas, and meaning. He goes on to identify the ways language can be used not only to illuminate but also to obfuscate, and he criticizes those who use dense prose to mask the weakness of their ideas. "Vague and insignificant forms of speech, and abuse of language, have so long passed for mysteries of science; and hard or misapplied words, with little or no meaning, have, by prescription, such a right to be mistaken for deep learning, and height of speculation, that it will not be easy to persuade . . . that they are but the covers of ignorance, and hinderance of true knowledge."[2] The danger Locke alludes to here is well known in the history of science. It is not difficult to find early modern treatises that use the language of natural philosophy to prop up superstitions or popular beliefs. This presented a challenge to authors who relied on a distinct set of empirical practices and invoked a specialized vocabulary—much of it technical. How could they invest a scientific treatise with authority when there were so many printed works vying for the same epistemological status? What could an author do about readers who failed to apprehend a work's meaning but who, nevertheless, levied criticisms? What about the intellectual thieves who appropriated ideas or discoveries and attempted to take credit for them? And how should a scientist or mathematician handle printers whose slow pace or (in)attention to detail could spell ruin for a publication? In the face of these and other hurdles, natural philosophers often became, as the title of this book suggests, *loath to print*.

Out of this reluctance a range of practices emerged, which I have explored in the preceding chapters. Some authors utilized the preface to the reader to rhetorically frame a work for its audience. While we certainly cannot take these prefaces at face value, surveying the tropes and themes found in the prefaces to scientific and medical works helps us understand how this paratextual element—a seemingly simple address to the person holding the book—was strategically composed to both guide the reader's experience and to establish a work's authority. Other times, authors attempted to control the readership for their books in more direct fashion, literally targeting readers through the careful distribution of copies. Whether it was Descartes' solicitation of responses from specific Sorbonne faculty member or Huygens' efforts to ensure that certain Royal Society members received his horology treatise, the practice of an author distributing copies of their work to particular readers reflects, in part, a desire to cultivate a much narrower audience than what print could naturally afford. A similar end might be achieved if an author could print a work themselves, producing just enough copies to share with colleagues. Such a scenario inspired several scientists to develop their own printing techniques. The results were not just clever instruments like the pantograph but also methods of reproducing images in more accurate or representative ways. John Evelyn, referring to engravings, noted that the sciences could be taught "by cuts alone." His history of printmaking, *Sculptura*, is a testimony to the importance of image reproduction in early modern science. And finally, the editor emerged in the publication of scientific, medical, and mathematical works in this period. On first pass, it is hard to see the editorial role as a response to the challenges of print culture, but on closer inspection it becomes evident that scientists who might otherwise have passed on publishing their work—for any of the aforementioned reasons—found a path to publication with the assistance of an editor.

Taken individually, each chapter in this book seems to address an isolated aspect of scientific printing in the early modern era, but collectively I believe that they show how a reluctance to print was managed and how frustrations and concerns were to some degree allayed when authors found alternative paths to publication. And on these paths, scientific authors often found new ways to bend print to fit their needs. Reluctance, at times, gave rise to innovation. The fields of print history and the history of science have long cross-pollinated. This study continues that trend by showing how natural philosophers responded to the prospect of printing their work and how they found creative ways to use print to their advantage.

Notes

INTRODUCTION: "A Vast Ocean of Books"

1. For clarity I use the generic (and admittedly anachronistic) term *science* to identify a broad category of knowledge, rather than the more nuanced disciplinary labels. In subsequent chapters the labels of natural philosopher, anatomist, etc. will emerge, but only in the context of specific examples. Broader conclusions will be couched in the more accessible terms of *science* and *scientific*. In doing this I draw on the rationale provided by Deborah Harkness in her prefatory note to *The Jewell House: Elizabethan London and the Scientific Revolution* (New Haven, CT: Yale University Press, 2007), xv–xviii.

2. Menso Folkerts, "Regiomontanus' Role in the Transmission and Transformation of Greek Mathematics," in *Tradition, Transmission, Transformation: Proceedings of Two Conferences on Pre-modern Science Held at the University of Oklahoma*, ed. F. Jamil Ragep and Sally P. Ragep (Leiden: E. J. Brill, 1996): 89–114.

3. Quoted in E. Zinner, *Regiomontanus: His Life and Works*, trans. E. Brown (Amsterdam: Elsevier Science, 1990), 108. The letter was from Hermann Schedel to his nephew Hartmann Schedel.

4. Zinner, *Regiomontanus*, 112–13.

5. Michael Shank, "The Geometrical Diagrams in Regiomontanus's Edition of His Own *Disputationes* (c. 1475): Background, Production, and Diffusion," *Journal for the History of Astronomy* 43 (2012): 29.

6. From Regiomontanus, *Disputationes contra Cremonensia in Planetarum Theoricas Delyramenta* (Nuremberg, 1474), quoted in Adam Mosely, *Bearing the Heavens: Tycho Brahe and the Astronomical Community of the Late Sixteenth Century* (Cambridge: Cambridge University Press, 2011), 149.

7. Rosamond McKitterick, "Books and the Sciences before Print," in *Books and the Sciences in History*, ed. Marina Frasca-Speda and Nick Jardine (Cambridge: Cambridge University Press, 2000), 13–34.

8. See Harold Love, *Scribal Publication in Seventeenth Century England* (Oxford: Clarendon Press, 1993); Julia Boffrey, *Manuscript and Print in London, c. 1475–1530* (London: British Library, 2012); and Richard Yeo, *Notebooks, English Virtuosi, and Early Modern Science* (Chicago: University of Chicago Press, 2014). On orality in the age of print, see Eileen Reeves, "Speaking of Sunspots: Oral Culture in an Early Modern Scientific Exchange," *Configurations* 13, no. 2 (Spring 2005): 185–210.

9. Elizabeth Eisenstein, *The Printing Revolution in Early Modern Europe* (Cambridge: Cambridge University Press, 2005): 13–17.

10. Ian MacLean, *Learning and the Market Place: Essays in the History of the Early Modern Book* (Leiden: Brill, 2009), 12, 65.

11. Owen Gingerich, *The Book Nobody Read: Chasing the Revolutions of Nicolaus Copernicus* (New York: Penguin, 2005), 126–28.

12. Lorenzo Perilli, "A Risky Enterprise: The Aldine Edition of Galen, the Failures of the Editors, and the Shadow of Erasmus of Rotterdam," *Early Science and Medicine* 17 (2012): 450.

13. Elizabeth Eisenstein, *The Printing Press as an Agent of Change* (Cambridge: Cambridge University Press, 1980).

14. See David McKitterick, *Print, Manuscript and the Search for Order* (Cambridge: Cambridge University Press, 2005); and Adrian Johns, *The Nature of the Book* (Chicago: University of Chicago Press, 1998).

15. David Wootton, *The Invention of Science: A New History of the Scientific Revolution* (New York: HarperCollins, 2015), 198.

16. Johannes Kepler, *De Stella Nova* (1606), cited in Adam Mosely, "Astronomical Books and Courtly Communication," in Frasca-Speda and Jardine, *Books and the Sciences in History*, 114.

17. Owen Gingerich and Robert S. Westman, "The Wittich Connection: Conflict and Priority in Late-Sixteenth Century Cosmology," *Transactions of the American Philosophical Society* 78, no. 7 (1988): 1–4.

18. Francis Bacon, *Works*, James Spedding et al. (London: Longman and Co., 1857), vol. 4, 271.

19. Edmund Spenser, *The Faerie Queen: The Shepheards Calendar: Together with the Other Works of England's Arch-poet* (London, 1617). On the frontispiece see Adam G. Hooks, "Sidney's porcupine," on ANCHORA, 2018, http://www.adamghooks.net/2011/04/sidneys -porcupine.html. There is some discussion as to whether the bush is rosemary or marjoram. In his *Colloquies* Erasmus notes, "The Marjorams word is *Abstine Sus, non tibi spiro: My Perfume was never made for the Snout of a Sow*; being a Fragrancy to which the Sow has a natural Aversion. And so every other herb has something in the Title, to denote the particular Virtue of the Plant." Desiderius Erasmus, *Twenty-two Select Colloquies out of Erasmus Roterodamus* [. . .] (London, 1689).

20. Michael Saenger, *The Commodification of Textual Engagements in the English Renaissance* (Aldershot: Ashgate, 2006), 48. See also Michael Mack, *Sidney's Poetics: Imitating Creation* (Washington, DC: Catholic University of America Press), 50–52; and Alfred W. Pollard, *Last Words on the History of the Title-Page with Notes on Some Colophons* [. . .] (London: John C. Nimmo, 1891).

21. Newton to Oldenburg, November 18, 1676, *The Correspondence of Isaac Newton*, ed. H. W. Turnbull (Cambridge: Cambridge University Press, 1959–1975), vol. 2, 183.

22. Johannes Kepler, preface to the reader, *Astronomia Nova* (1609).

23. William Gilbert, "To the Candid Reader," *De Magnete* (London, 1600), sig. iij, verso.

24. Huygens to Fatio de Duillier, January 28, 1689/90, in Turnbull, *Correspondence*, vol. 3, 67.

25. Leibniz to Newton, March 7, 1692/93, in Turnbull, *Correspondence*, vol. 3, 258.

26. In this case, "large" is a relative term. Comparing incunable production to manu-

script copying, print represents a quantum leap, with hundreds of copies produced in even the earliest print runs. Runs under one hundred copies were rare and hardly worthwhile, financially. See Jonathan Greene, Frank McIntyre, and Paul Needham, "The Shape of Incunable Survival and Statistical Estimation of Lost Editions," *Papers of the Bibliographical Society of America* 105, no. 2 (June 2011): 144.

27. Adrian Johns, *The Nature of the Book: Print and Knowledge in the Making* (Chicago: University of Chicago Press, 1998).

CHAPTER ONE: **Authorial Attitudes toward Print**

1. William Gilbert, "To the Candid Reader," *De Magnete* (London, 1600), sig. iij, verso.

2. John Locke, epistle to the reader, *An Essay concerning Human Understanding* (London, 1689), xi.

3. Anne Goldgar, *Impolite Learning: Conduct and Community in the Republic of Letters, 1680–1750* (New Haven, CT: Yale University Press, 1995).

4. Philip Gaskell, *A New Introduction to Bibliography* (New Castle, DE: Oak Knoll Press, 1995).

5. Lucien Febvre and Henri-Jean Martin, *The Coming of the Book*, 3rd ed. (London: Verso, 2010).

6. The exact number of works printed from 1455 to 1500 is a challenge to determine because of limited print run data. See Eric Marshall White, "A Census for Print Runs of Fifteenth Century Books," Consortium of European Research Libraries, accessed October 16, 2016, https://www.cerl.org/_media/resources/links_to_other_resources/printruns _intro.pdf. The story of print's spread, however rapid, has been given nuance by Neil Harris, who effectively argues that many traveling German printers were itinerant because their product was less than successful. "In some seventy Italian centers, some of them no larger than villages, printing appeared momentarily, at times more than once, in the fifteenth century and subsequently disappeared for fifty, seventy, sometimes a hundred years." Harris, "The Italian Renaissance Book: Catalogues, Censuses and Survival," in *Book Triumphant: Print in Transition in the Sixteenth and Seventeenth Centuries*, ed. Malcolm Walsby and Graeme Kemp (Leiden: Brill, 2011), 28.

7. Data based on the Incunable Short Title Catalogue, British Library, accessed October 16, 2016, http://www.bl.uk/catalogues/istc/.

8. Elizabeth Eisenstein, *The Printing Press as an Agent of Change* (Cambridge: Cambridge University Press, 1980), chapter 2.

9. Henri-Jean Martin, *Print, Power, and People in 17th-Century France*, trans. David Gerard (Metuchen, NJ: Scarecrow Press, 1993).

10. Martin, *Print, Power, and People*, 157.

11. The lunar map published by Michael Florent van Langren in 1645 is a good example of this kind of astronomical ephemera; it was a broadsheet that illustrated the moon's surface and bore the title *Lumina Austriaca Philippica*. There are two known copies of Blaise Pascal's *Essay pour les coniques* broadside, one held in the Bibliothèque nationale in Paris and the other in Leibniz's archived papers in Hanover. On this broadside, see Frances Marguerite Clarke and David Eugene Smith, "*Essay pour les Coniques* of Blaise Pascal," *Isis* 10, no. 1 (March 1928): 16–20.

12. "Vsvs tractatio gnomonis magni . . . with diagram of a Sundial," Bologna, 1576, Museum of the History of Science, Oxford University, accessed September 21, 2018, http://www.mhs.ox.ac.uk/object/inv/14044.

13. Adrian Johns, *The Nature of the Book* (Chicago: Chicago University Press, 1998), 36.

14. Johns, *The Nature of the Book*, 36. Johns' focus is on this indeterminacy and the way that book history, as a field and an intellectual tool, can tell us how individuals in the period navigated the epistemic weakness in order to assert their ideas on firmer ground.

15. Cited in Adam Mosley, *Bearing the Heavens: Tycho Brahe and the Astronomical Community of the Late Sixteenth Century* (Cambridge: Cambridge University Press, 2007), 149.

16. Mario Biagioli, "From Ciphers to Confidentiality: Secrecy, Openness and Priority in Science," *British Journal for the History of Science* 45, no. 2 (June 2012): 214.

17. Quoted in Harold Love, *Scribal Publication in the Seventeenth-Century* (Amherst: University of Massachusetts Press, 1993), 153.

18. D. F. McKenzie, "Speech-Manuscript-Print," in *New Directions in Textual Studies*, ed. Dave Oliphant and Robin Bradford (Austin: University of Texas Press, 1990), 98.

19. D. F. McKenzie, Peter McDonald, and Michael Suarez, *Making Meaning: "Printers of the Mind" and Other Essays* (Amherst: University of Massachusetts Press, 2002), 238.

20. Lister to Oldenburg, in *The Correspondence of Henry Oldenburg*, ed. A. Rupert Hall and Marie Boas Hall (Madison: University of Wisconsin Press, 1969), #2081, vol. 9, p. 282. Page numbers for subsequent quotes are cited parenthetically as *CHO*.

21. Newton to Oldenburg, in *The Correspondence of Isaac Newton*, ed. H. W. Turnbull et al. (Cambridge: Cambridge University Press, 1959–1977), vol. 1, #42. Page numbers for subsequent quotes are cited parenthetically as *CIN*.

22. Adrian Johns, "The Ambivalence of Authorship in Early Modern Natural Philosophy," in *Scientific Authorship: Credit and Intellectual Property in Science*, ed. Mario Biagioli and Peter Galison (New York: Routledge, 2003), 68.

23. Andrew Pettegree, *The Book in the Renaissance* (New Haven, CT: Yale University Press, 2010), 205. See also Gigliola Fragnito, ed., *Church, Censorship and Culture in Early Modern Italy*, trans. Adrian Belton (Cambridge: Cambridge University Press, 2001).

24. Pettegree, *The Book in the Renaissance*, 206.

25. Descartes to Mersenne, November 1634, *The Philosophical Writings of Descartes*, trans. and ed. John Cottingham et al. (Cambridge: Cambridge University Press, 1991), 3:41 (emphasis added).

26. Descartes to Mersenne, February 1634, *Philosophical Writings*, 42.

27. Stephen Gaukroger, *Descartes: An Intellectual Biography* (Oxford: Clarendon Press, 1995), 187. On censorship in the Low Countries, see Joris van Eijnatten, "Between Practice and Principle: Dutch Ideas on Censorship and Press Freedom, 1579–1795," *Redescriptions: Political Thought, Conceptual History and Feminist Theory* 8, no. 1 (January 2004): 85–113.

28. Horace, *Ars Poetica*, trans. H. R. Fairclough (Cambridge, MA: Harvard University Press, 1926), 483.

29. Howard Jones, *The Epicurean Tradition* (London: Routledge, 1989), 168.

30. Huygens to Oldenburg, June 16, 1669, *CHO*, vol. 4, 45.

31. Leibniz to Oldenburg, April 29, 1671, *CHO*, vol. 8, 28.

32. Chloe Wheatley, *Epic, Epitome, and the Early Modern Historical Imagination* (Sur-

rey: Ashgate, 2011), 13. Epitomizing was closely linked to theories of taxonomy in knowledge production, with authors such as Francis Bacon writing about how best to organize the information gained through reading. See Richard Yeo, *Notebooks, English Virtuosi, and Early Modern Science* (Chicago: University of Chicago Press, 2014).

33. John Wallis to John Collins, November 14, 1672, *The Correspondence of Scientific Men of the Seventeenth Century, Including Letters of Barrow, Flamsteed, Wallis, and Newton*, ed. Stephen Peter Rigaud and Stephen Jordan Rigaud (Oxford: Oxford University Press, 1841), vol. 2, 552. It is worth noting that epitomes and abridgments are not precisely the same.

34. See Stephen Gaukroger, *Francis Bacon and the Transformation of Early-Modern Philosophy* (Cambridge: Cambridge University Press, 2001), chapter 1, pp. 9–11.

35. Gaukroger, *Francis Bacon*, 9.

36. Jürgen Habermas, *The Structural Transformation of the Public Sphere: An Inquiry into a Category of Bourgeois Society*, trans. Thomas Burger and Frederick Lawrence (Cambridge, MA: MIT Press, 1991).

37. G. W. F. Hegel, *Elements of the Philosophy of Right*, ed. Allen W. Wood, trans. H. B. Nisbet (Cambridge: Cambridge University Press, 1991), §318:357–58.

38. David Zaret, "Religion, Science, and Printing in the Public Spheres in Seventeenth-Century England," in *Habermas and the Public Sphere*, ed. Craig Calhoun (Cambridge, MA: MIT Press, 1992), chapter 9.

39. Peter Lake and Steve Pincus, "Rethinking the Public Sphere in Early Modern England," *Journal of British Studies* 45, no. 2 (April 2006): 270–92. See also Jan C. Rupp, "The New Science and the Public Sphere in the Premodern Era," *Science in Context* 8, no. 3 (1995): 487–507.

40. See Steven Shapin and Simon Schaffer's *Leviathan and the Air Pump: Hobbes, Boyle, and the Experimental Life* (Princeton, NJ: Princeton University Press, 1985).

41. Galileo Galilei, "Letter to the Grand Duchess Christina," in *Discoveries and Opinions of Galileo*, trans. Stillman Drake (New York: Anchor Books, 1957), 189.

42. Galileo, "Letter to the Grand Duchess Christina," 190.

43. Jean Dietz Moss, "Galileo's Letter to Christina: Some Rhetorical Considerations," *Renaissance Quarterly* 36, no. 4 (Winter 1983): 552.

44. Galileo, "Letter to the Grand Duchess Christina," 176.

45. Meredith K. Ray, *Daughters of Alchemy: Women and Scientific Culture in Early Modern Italy* (Cambridge, MA: Harvard University Press, 2015), 120. Erculiani's *Letters on Natural Philosophy* contained a preface to the reader which, like many others discussed here, adopted the tropes of scientific prefaces of the period, including a reluctance to publish, a concern for piracy of her ideas, and a capitulation to the urges of an unnamed friend who insisted on seeing her work in print.

46. Ray, *Daughters of Alchemy*, 120–21.

47. Ann Blair, *Too Much to Know: Managing Scholarly Information before the Modern Age* (New Haven, CT: Yale University Press, 2010), 55.

48. Rietje van Vliet, "Print and Public in Europe, 1600–1800," in *A Companion to the History of the Book*, ed. Simon Eliot and Jonathan Rose (West Sussex: Wiley-Blackwell, 2009), 248.

49. Robert Burton, *The Anatomy of Melancholy* (New York: New York Review of Books, 2001), 18 and 22. On Burton's attitude specifically, see David J. Baker, "Robert Burton and the Discontents of Print," in *A Companion to British Literature*, vol. 2: *Early Modern Literature 1450–1660*, ed. Robert DeMaria et al. (West Sussex: John Wiley & Sons, 2014), 29–39.

50. Ann Blair, *Too Much to Know: Managing Scholarly Information before the Modern Age* (New Haven, CT: Yale University Press, 2010), 58.

51. William Gilbert, preface, *De Magnete* (London, 1600), sig. iij.

52. John Pell, *Sir: The summe of what I have heretofore written or spoken to you, concerning the advancement of the mathematickes, is this: as long as men want will, wit, meanes or leisure to atte[nd] those studies, it is no marvaile if they make no great progresse in them. To remedy which, I conceive these meanes not to be amisse* (London, 1638).

53. Note that italics and brackets in the quotation are original and not added.

54. Pell, *Sir*.

55. Philip P. Wiener, ed., *Leibniz Selections* (New York: Charles Scribner's Sons, 1951), 29.

56. Wiener, *Leibniz Selections*, 29–32.

57. Robert Clavel, *A Catalogue of All the Books Printed in England Since the Dreadful Fire of London, in 1666: To the End of Michaelmas Term, 1672* (London, 1673), 24.

58. Edward Arber, ed., *The Term Catalogues, 1668–1709*, vol. 2: *May 1683* (West Kensington, 1905), 18.

59. Boyle to Samuel Hartlib, September 14, 1655, in *Correspondence of Robert Boyle*, ed. Michael Hunter, Antonio Clericuzio, and Laurence Principe (London: Pickering & Chatto, 2001), vol. 1, 190–91.

60. Timothy Clarke to Oldenburg, April/May 1668, *CHO*, vol. 4, 362–63.

61. Clarke to Oldenburg, April/May 1668, 363.

62. Barnabe Rich, *Opinion diefied [sic]: Discovering the ingins, traps, and traynes, that are set in this age, whereby to catch opinion* (London: printed for Thomas Adams, 1613), 2.

63. Descartes to Mersenne, October 8, 1629, *The Philosophical Writings of Descartes*, trans. and ed. John Cottingham et al. (Cambridge: Cambridge University Press, 1991), vol. 3, 6–7.

64. In the end, he published the work in French, but that decision was made at a later date.

65. Mary Terrall, "The Uses of Anonymity in the Age of Reason," in *Scientific Authorship: Credit and Intellectual Property in Science*, ed. Mario Biagioli and Peter Galison (New York: Routledge, 2003), 91 and chapter 4. See also David Kronick, "Anonymity and Identity: Editorial Policy in the Early Scientific Journal," in *"Devant le Deluge" and Other Essays on Early Modern Scientific Communication* (Oxford: Scarecrow Press, 2004), 136–52.

66. Descartes to Mersenne, April 15, 1630, *Philosophical Writings*, 20–21.

67. Descartes to Mersenne, November 25, 1630, *Philosophical Writings*, 28.

68. Descartes to Mersenne, February 27, 1637, *Philosophical Writings*, 54.

69. Justel to Oldenburg, September 1668, *CHO*, vol. 5, 39.

70. Descartes to Mersenne, March 1636, *Philosophical Writings*, vol. 3, 51.

71. Descartes to Mersenne, February 1637, *Philosophical Writings*, vol. 3, 53.

72. Descartes to Vatier, February 22, 1638, *Philosophical Writings*, vol. 3, 86.

73. Erica Harth, *Cartesian Women: Versions and Subversions of Rational Discourse in the Old Regime* (Ithaca, NY: Cornell University Press, 1992), 69–72.

74. David McKitterick, *A History of Cambridge University Press* (Cambridge: Cambridge University Press, 1992), vol. 1, 238–39.

75. Descartes to Huygens, October 5, 1637, *Philosophical Writings*, 66.

76. Cited in Desmond M. Clark, *Descartes* (Cambridge: Cambridge University Press, 2006), 138.

77. Mordechai Feingold, ed., *Before Newton: The Life and Times of Isaac Barrow* (Cambridge: Cambridge University Press, 1990), 71–72.

78. Collins to Gregory, December 24, 1670, *CIN*, vol. 1, 55.

79. Rob Iliffe, "Butter for Parsnips," in *Scientific Authorship: Credit and Intellectual Property in Science*, ed. Mario Biagioli and Peter Galison (London: Routledge, 2003), 43.

80. Newton to Oldenburg, November 18, 1676, *CIN*, vol. 2, 182.

81. Collins to Gregory, December 24, 1670.

82. See Katherine Hunt, "Convenient Characters: Numerical Tables in William Godbid's Printed Books," *Journal of the Northern Renaissance* 6 (2014), http://www.northern renaissance.org/convenient-characters-numerical-tables-in-william-godbids-printed -books/; and Henry R. Plomer, *A Dictionary of the Booksellers and Printers Who Were at Work in England, Scotland and Ireland from 1641–1667* (London: Bibliographical Society, 1907), 83.

83. Spinoza to Oldenburg, April 1662, *CHO*, vol. 1, 466. This manuscript would ultimately be published posthumously in 1677 as *Treatise on the Improvement of the Understanding*.

84. Ibid., 472–73.

85. Oldenburg to Spinoza, April 3, 1663, *CHO* vol. 2, 43.

86. Harold Love, *The Culture and Commerce of Texts: Scribal Publication in Seventeenth-Century England* (Amherst: University of Massachusetts Press, 1993).

87. David McKitterick, *Print, Manuscript and the Search for Order, 1450–1830* (Cambridge: Cambridge University Press, 2003), 206.

88. Michael Hunter, "Robert Boyle and Secrecy," in *Secrets and Knowledge in Medicine and Science, 1500–1800*, ed. Elaine Leong and Alisha Rankin (New York: Routledge, 2016), 87–104.

89. Robert Boyle, *Certain physiological essays and other tracts* (London, 1669), 123.

90. Michael Hunter, *Boyle Studies: Aspects on the Life and Thought of Robert Boyle (1627–91)* (Surrey: Ashgate, 2015), 156. See also Harriet Knight and Michael Hunter, "Robert Boyle's *Memoirs for the Natural History of Human Blood* (1684): Print, Manuscript and the Impact of Baconianism in Seventeenth-Century Medical Science," *Medical History* 51 (2007): 145–64.

91. Hunter, *Boyle Studies*, 157.

92. Lindsay O'Neill, *The Opened Letter: Networking in the Early Modern British World* (Philadelphia: University of Pennsylvania Press, 2015), 140–41. See also James Daybell and Andrew Gordon, eds., *Cultures of Correspondence in Early Modern Britain* (Philadelphia: University of Pennsylvania Press, 2016); Francisco Bethencourt and Florike Egmond,

eds., *Cultural Exchange in Early Modern Europe*, vol. 3: *Correspondence and Cultural Exchange in Europe, 1400–1700* (Cambridge: Cambridge University Press, 2007); and Dena Goodman, *The Republic of Letters: A Cultural History of the French Enlightenment* (Ithaca, NY: Cornell University Press, 1994).

93. Michael Hunter, *Establishing the New Science: The Experience of the Early Royal Society* (Woodbridge, Suffolk: Boydell Press, 1989), 253.

94. John Beale, *Herefordshire Orchards, A Pattern For all England. Written in an Epistolary Address to Samuel Hartlib, Esq.* (London, 1657), 59–60.

95. Peter N. Miller, *Peiresc's Mediterranean World* (Cambridge, MA: Harvard University Press, 2015), 15.

96. Miller, *Peiresc's Mediterranean World*, 17.

97. See Brian Ogilvie, "Correspondence Networks," in *A Companion to the History of Science*, ed. Bernard V. Lightman (Malden, MA: John Wiley & Sons, 2016), 358–71. On Oldenburg specifically, see Maurizio Gotti, "Scientific Interaction within Henry Oldenburg's Letter Network," *Journal of Early Modern Studies* 3 (2014): 151–71. See also Marie Boas Hall, *Henry Oldenburg: Shaping the Royal Society* (Oxford: Oxford University Press, 2002), chapter 5. On Mersenne, see Justin Grosslight, "Small Skills, Big Networks: Marin Mersenne as Mathematical Intelligencer," *History of Science* 51, no. 3 (September 2013): 337–74.

98. British Library Add. MS 4384. Accessed through EEBO, February 2, 2017.

99. British Library Add. MS 4384, image 39 of 61.

100. Beale to Oldenburg, April 1, 1664, *CHO*, vol. 2, 161.

101. Justin Grosslight, "Small Skills, Big Networks," 348.

102. Auzout to Oldenburg, September 23, 1665, *CHO*, vol. 2, 518.

103. Huygens to J. Chapelain, March 28, 1658, in *Oeuvres Complètes de Christiaan Huygens* (La Haye: M. Nijhoff, 1888–1950), vol. 2, 157. Hereafter cited as *OCCH*.

104. See Anne Goldgar, *Impolite Learning: Conduct and Community in the Republic of Letters, 1680–1750* (New Haven, CT: Yale University Press, 1995).

105. The quote comes from Jacqueline Stedall, "John Wallis and the French: His Quarrels with Fermat, Pascal, Dulaurens and Descartes," *Historia Mathematica* 39, no. 3 (August 2012): 272.

106. Wallis to Oldenburg, March 30, 1668, *CHO*, vol. 4, 285–86.

107. Justel to Oldenburg, May 27, 1668, *CHO*, vol. 4, 492.

108. Justel to Oldenburg, 10 June 1668, *CHO*, vol. 4, 462.

109. "Wallis's Animadversions on Dulaurens," *CHO*, vol. 4, 494 (emphasis added).

110. Wallis's Animadversions on Dulaurens," 494.

111. Catherine Goldstein, "Routine Controversies: Mathematical Challenges in Mersenne's Correspondence," arXiv.org, September 24, 2012, https://arxiv.org/abs/1209.5376. See also Goldstein, "Neither Public nor Private: Mathematics in Early Modern France," accessed December 12, 2016, https://webusers.imj-prg.fr/~catherine.goldstein/Places -Goldstein.pdf.

112. Michael Mahoney, *The Mathematical Career of Pierre de Fermat, 1601–1665*, 2nd ed. (Princeton, NJ: Princeton University Press, 1994), 23.

113. Mahoney, *The Mathematical Career of Pierre de Fermat*, 344.

114. Quoted in Catherine Goldstein, "Neither Public nor Private," 10. Goldstein goes on to consider these mathematical challenges in the context of gender, a study well worth examining.

115. Collins to Gregory, December 24, 1670, *CIN*, vol. 1, 57–58.

116. *Correspondence of John Wallis (1616–1703)*, ed. Philip Beeley and Christoph J. Scriba (Oxford: Oxford University Press, 2014), vol. 4, 166–67.

117. Thomas Birch, *The History of the Royal Society of London* (London: printed for Al. Millar, 1758), vol. 4, 223.

118. Quoted in Noel Malcolm, *Aspects of Hobbes* (Oxford: Oxford University Press, 2002), 164–65.

119. *The Correspondence of Thomas Hobbes*, ed. Noel Malcolm (Oxford, Oxford University Press, 1994), vol. 2, 906.

120. See Mario Biagioli, "From Ciphers to Confidentiality: Secrecy, Openness and Priority in Science," *British Journal for the History of Science* 45, no. 2 (June 2012): 213–33.

121. Stillman Drake, "Galileo, Kepler and the Phases of Venus," *Journal of the History of Astronomy* 15, no. 3 (October 1984): 201.

122. *OCCH*, vol. 15, 167–69. See also Huygens to Wallis, June 13, 1655, vol. 1 p. 332.

123. Albert van Helden, "The Telescope in the Seventeenth Century," *Isis* 65 (1974): 44; van Helden, "Annulo Cingitur: The Solution of the Problem of Saturn," *Journal for the History of Astronomy* 5 (1974): 155–74.

124. Wallis to Huygens, March 22, 1655/56, in Beeley and Scriba, *Correspondence*, vol. 1, 178.

125. *OCCH*, vol. 15, 177.

126. *OCCH*, vol. 15, 177. "Annulo cingitur tenui, plano, nusquam cohaerent, ad eclipticam inclinato."

127. See Mario Biagioli, *Galileo's Instruments of Credit* (Chicago, University of Chicago Press, 2006).

128. Mario Biagioli, "From Prints to Patents: Living on Instruments in Early Modern Europe," *History of Science* 44 (2006): 139–86.

129. Lisa Jardine, "Monuments and Microscopes: Scientific Thinking on a Grand Scale in the Early Royal Society," *Notes and Records of the Royal Society of London* 55, no. 2 (May 2001): 296.

130. Allan Chapman, *England's Leonardo: Robert Hooke and the Seventeenth-Century Scientific Revolution* (Bristol: Institute of Physics, 2005), 174.

131. Robert Hooke, *A Description of Helioscopes and Some Other Instruments* (London, 1676), 32. The anagram, composed of two distinct phrases, was aaaæbccddeeeeeegiiilmmm- nnooppqrrrrstttuuuuu / aaeffhiiiillnrrsstuu. Stephen Inwood discusses this anagram and its relation to steam power in *The Forgotten Genius: The Biography of Robert Hooke, 1635– 1703* (San Francisco: MacAdam/Cage, 2003), 198.

132. Newton to Oldenburg, October 24, 1676, *CIN*, vol. 2, 152.

133. Newton to Oldenburg, October 24, 1676, 153. The anagram's solution was *aequatione quotcunque fluentes quantitates involvente, fluxiones invenire, et vice versa* which translates as "given an equation involving any number of fluent quantities to find the fluxions, and conversely."

134. Newton to Oldenburg, October 26, 1676, *CIN* vol. 2, 162–63.

135. Newton to Collins, November 8, 1676, *CIN* vol. 2, 179., *CIN*, 179.

136. Wallis to Newton, April 10, 1695, *Correspondence of Sir Isaac Newton and Professor Cotes*, ed. J. Edelston (London: John W. Parker, 1850), 300.

137. Collins to Newton, August 30, 1677, *CIN*, vol. 2, 237.

138. Collins to Baker, February 10, 1676/77, *CIN*, vol. 2, 192.

139. Newton to Oldenburg, November 18, 1676, *CIN*, vol. 2, 183.

CHAPTER TWO: "To the Unprejudiced Reader"

1. Newton to Oldenburg, February 10, 1671, *The Correspondence of Isaac Newton*, ed. H. W. Turnbull et al. (Cambridge: Cambridge University Press, 1959–1977).

2. On Huygens, see C. D. Andriesse, *Huygens: The Man Behind the Principle*, trans. Sally Miedema (Cambridge: Cambridge University Press, 2011); and Joella G. Yoder, *Unrolling Time: Christiaan Huygens and the Mathematization of Nature* (Cambridge: Cambridge University Press, 2004).

3. Christiaan Huygens, *The Celestial Worlds Discovered; or, Conjectures Concerning the Inhabitants, Plants and Productions of the Worlds in the Planets*, 2nd ed. (London: James Knapton, 1722), 5.

4. These examples taken from the following texts, respectively: William Bullein, *Bulwarke of Defence against All Sickness* (London: Thomas Marshe, 1579); Henry Cornelius Agrippa, *His Fourth Book of Occult Philosophy*, trans. Robert Turner (London: printed by J. C. for Thos. Rooks, 1665); Christopher Merret, *The Accomplisht Physician, The Honest Apothecary, and the Skilful Chyrurgeon* (London, 1670).

5. Merrett, "Preface to the Vulgar Reader," in *The Accomplisht Physician, The Honest Apothecary, and the Skillful Chyrurgeon*, sig. A2 (emphasis mine).

6. On Gideon Harvey as author of this work, see Lynda Payne, *With Words and Knives: Learning Medical Dispassion in Early Modern England* (Hampshire: Ashgate, 2013), 30n79. See also H. A. Colwell, "Gideon Harvey: Sidelights on Medical Life from the Restoration to the End of the XVII Century," *Annals of Medical History* 3, no. 3 (Fall 1921): 233. On G. Harvey more generally, see Roger French and Andrew Wear, eds., *The Medical Revolution of the Seventeenth Century* (Cambridge: Cambridge University Press, 1989), 187–88.

7. Despite these beliefs, Gideon Harvey was appointed physician-in-ordinary to Charles II and subsequently physician to William and Mary. See Patrick Wallis, "Gideon Harvey," in *The Oxford Dictionary of National Biography* (September 2004).

8. Gerard Genette, *Paratexts: Thresholds of Interpretation* (Cambridge: Cambridge University Press, 1997), 1–2. Genette actually differentiates between paratext that sits within or alongside a text—what he calls peritext—and paratext that comes at the end of a work, such as the index or colophon. The latter he calls epitext. For my purposes, I will use the more blanket term paratext when discussing prefaces.

9. Genette, *Paratexts*, 2 (emphasis original).

10. Authorial intention was initially deemed inaccessible by W. K. Wimsatt and M. C. Beardsley in "The Intentional Fallacy," *Sewanee Review* 54, no. 3 (July–Sept 1946): 468–88. In this chapter I presume that authorial prefaces are not to be read as literal expressions of

an author's intentions. Rather, they are authorial masks, signaling the ways an author hopes to position themselves. Prefaces, then, are postures, but no less informative for being so.

11. William H. Sherman, "Architecture, Paratext, and Early Print Culture," in *Agent of Change: Print Culture Studies after Elizabeth L. Eisenstein* (Amherst: University of Massachusetts Press, 2007), 79.

12. William Caxton, trans., epilogue to *The Recuyell of the Historyes of Troye*, ed. H. Oskar Sommer (London: David Nutt, 1894), book 3.

13. These figures were kindly shared by Meaghan Brown, Fellow for Data Curation in Early Modern Studies at the Folger Library. They are based on surveys of STC prefaces in the years described.

14. Johann Wecker, *A compendious chyrugerie* (London: imprinted by John Windet, 1585). "The Booke to the Reader" offers the following promises in what is the only preface I've found written from the perspective of the book itself: "I swellings wast, I wounds do joyne / I ulcers do make sounde: / I doe the broken bones restore, / what further can be founde?"

15. Jacques Derrida, *Dissemination*, trans. Barbara Johnson (London: Continuum, 1981), 7.

16. Walter J. Ong, "The Writer's Audience Is Always a Fiction," *PMLA* 90, no. 1 (January 1975): 18.

17. Genette, *Paratexts*, 197.

18. Robert Boyle, preface to the reader, in *New Experiments Physico-Mechanicall Touching the Spring of the Air* (Oxford: printed by H. Hall for Tho. Robinson, 1662). William Harvey, *Exercitatio Anatomica, De Motu Cordis et Sanguinis in Animalibus* (Frankfurt, 1628).

19. Boyle, preface to the reader, in *New Experiments*.

20. Kenelm Digby, preface to the reader, in *The Closet of the Eminently Learned Sir Kenelme Digbie Kt. Opened* (London, 1669).

21. Jonathan Swift, *Gulliver's Travels*, ed. Roberta A. Greenberg (New York: W. W. Norton, 1970), iv.

22. Swift, *Gulliver's Travels*, vi.

23. J. W. Saunders, "The Stigma of Print: A Note on the Social Bases of Tudor Poetry," *Essays in Criticism* 1 (1951): 139–64. See also Daniel Traister, "Reluctant Virgins: The Stigma of Print Revisited," *Colby Quarterly* 26, no. 2 (June 1990): 75–86.

24. Wendy Wall, *The Imprint of Gender: Authorship and Publication in the English Renaissance* (Ithaca, NY: Cornell University Press, 1993).

25. For a more thorough discussion of Di Strata's attitude toward print, see John Martin, *Venice's Hidden Enemies: Italian Heretics in a Renaissance City* (Berkeley: University of California Press, 1993), 79. Among Di Strata's major quarrels with the press was that it made information available to the ignorant, a concern expressed by many clerics but one shared by many natural philosophers as well.

26. Wall, *The Imprint of Gender*, 156.

27. *The Philosophical Writings of Descartes*, ed. John Cottingham et al. (Cambridge: Cambridge University Press, 1985), vol. 1, 177.

28. Thomas Dekker, *A strange horse-race* (London, 1613), A3r. Here *Lectores* is Latin for "readers," while *Lictores* refers to the ancient Roman bodyguards of magistrates.

29. Kevin Dunn, *Pretexts of Authority: The Rhetoric of Authorship in the Renaissance Pref-ace* (Palo Alto: Stanford University Press, 1994, 15.

30. Dunn, *Pretexts of Authority*, 123.

31. Martin Elsky, *Authorizing Words* (Ithaca, NY: Cornell University Press, 1989), chapter 6.

32. Elsky, *Authorizing Words*, 186.

33. Francis Bacon, *Advancement of Learning*, ed. William Aldis Wright (Oxford: Clarendon Press, 1876), book 1, §6.

34. See Mario Biagioli, *Galileo Courtier* (Chicago: Chicago University Press, 1993); and Biagioli, *Galileo's Instruments of Credit* (Chicago: University Chicago Press, 2007).

35. See Mario Biagilio and Peter Galison, eds., *Scientific Authorship: Credit and Intellectual Property in Science* (New York: Routledge, 2003).

36. See Roger Chartier, "Foucault's Chiasmus," in *Scientific Authorship: Credit and Intellectual Property in Science* (New York: Routledge, 2003), 20.

37. Giovanni Alfonse Borelli, *De Motu Animalium* (Rome, 1680). I consulted the 1743 edition published by Gosse in The Hague and available at https://archive.org/details/demo tuanimaliumoojesugoog (accessed June 2017).

38. James Voelkel, *The Composition of Kepler's "Astronomia Nova"* (Princeton, NJ: Princeton University Press, 2011), 1.

39. Johannes Kepler, *De Fundamentis Astrologiae Certioribus*, trans. J. V. Field in "A Lutheran Astrologer: Johannes Kepler," *Archive for the History of Exact Sciences* 31, no. 3 (1984): 230. Page numbers for subsequent quotes are cited parenthetically.

40. James Voelkel, *The Composition of Kepler's "Astronomia Nova,"* 223.

41. Richard Oosterhoff, "Idiotae, Mathematics, and Artisans: The Untutored Mind and the Discovery of Nature in the Fabrist Circle," *Intellectual History Review* 3 (July 2014): 301–19, http://dx.doi.org/10.1080/17496977.2014.891180.

42. Kepler, *Astronomia Nova*, 65–66.

43. Isaac Newton, *Principia*, translated by I. Bernard Cohen and Anne Whitman (Berkeley: University of California Press, 1999), preface to book 3.

44. Robert Boyle, "The Author's Advertisement," in *Some considerations touching the vsefulnesse of experimental naturall philosophy* (London, 1663).

45. John Wilkins, *The Discovery of a World in the Moone: Or, A Discourse Tending to Prove, that 'tis probable there may be another habitable world in that planet* (London: printed by E.G. for Michael Sparke and Edward Forrest, 1638).

46. Wilkins, preface to the reader, in *The Discovery of a World in the Moone*, sig. A3.

47. Wilkins, preface to the reader, in *The Discovery of a World in the Moone.*

48. Wilkins, preface to the reader, in *The Discovery of a World in the Moone*, 6.

49. See Bruce Janacek, *Alchemical Belief: Occultism in the Religious Culture of Early Modern England* (University Park: Penn State University Press, 2011), 47–48.

50. Kenelm Digby, "To My Sonne," in *Two Treatises in the One of Which, the Nature of Bodies, In the Other, the Nature of Man's Soule* [. . .] (Paris: printed by Gilles Blaizot, 1644), sig. A2–B.

51. Digby, preface to *Two Treatises*, sig. B2–B4.

52. René Descartes, *Meditations on First Philosophy*, ed. and trans. John Cottingham (Cambridge: Cambridge University Press, 2013), 13.

53. See Gary Hatfield, *Descartes and the "Meditations"* (New York: Routledge, 2014).

54. Descartes, *Philosophical Writings*, vol. 3, 153. On the *Meditations*, see Roger Ariew and Donald Cress, introduction to René Descartes, *Meditations, Objections, and Replies*, eds. and trans. Roger Ariew and Donald Cress (Indianapolis, IN: Hackett, 2006).

55. René Descartes, *The Meditations*, in *Philosophical Writings*, preface to vol. 2, 8.

56. Reneé Raphael, *Reading Galileo: Scribal Technologies and the "Two New Sciences"* (Baltimore: Johns Hopkins University Press, 2017), 25.

57. Dunn, *Pretexts of Authority*, 101. It is interesting that although Dunn thinks Descartes was never ultimately successful in reconciling his public and private identities, he looks to Francis Bacon as an exemplar of the authorial role that was created in the late Renaissance, someone who fully embraced the identity of author and aristocrat.

58. In letters to various individuals, Descartes talks about epistolary exchanges as conversations akin to face-to-face discussions. He also frames the *Meditations* as conversations, describing written replies and objections in the same spirit as verbalized academic debates with the traditional back-and-forth of viewpoints. "If they [theologians and others] want to comment further on my replies, I will be very happy to produce my own additional comments." Descartes to Mersenne, March 4, 1641, vol. 3, p. 173.

59. René Descartes, *Principles of Philosophy*, in *Philosophical Writings*, vol. 1, 185.

60. Descartes, *Principles of Philosophy*, 185.

61. William H. Sherman, "What Did Renaissance Readers Write in Their Books?," in *Books and Readers in Early Modern England: Material Studies*, ed. Jennifer Andersen and Elizabeth Sauer (Philadelphia: University of Pennsylvania Press, 2002): 124.

62. Descartes, *Principles of Philosophy*, 185.

63. Descartes, *Principles of Philosophy*, 189.

64. Descartes, *Principles of Philosophy*, 189.

65. Robert Boyle, "The Author's Advertisement," in *Some considerations touching the usefulnesse of experimental natural philosophy* (London, 1663), n.p. See also John T. Harwood, "Science Writing and Writing Science: Boyle and Rhetorical Theory," in *Robert Boyle Reconsidered*, ed. Michael Hunter (Cambridge: Cambridge University Press, 1994), 48–50.

66. Michael Hunter, *Boyle Studies: Aspects of the Life and Thought of Robert Boyle (1627–91)* (Surrey: Ashgate, 2015), 159.

67. Margaret Cavendish, *The Worlds Olio* [. . .] (London, 1655). In this work Cavendish presents "An Epistle to the Reader," "The Preface to the Reader," and "To the Reader," the latter being a more traditional address to the audience. On Cavendish's appearance at the Royal Society, fellow John Evelyn writes:

Her head-gear was so pretty
I ne'ere saw anything so witty
Tho I was halfe a feard
God blesse us when
I first did see her

She look'd so like a cavaliere
But that she had no beard.

Quoted from Lisa T. Sarasohn, *The Natural Philosophy of Margaret Cavendish: Reason and Fancy during the Scientific Revolution* (Baltimore: Johns Hopkins University Press, 2010), 26.

68. Cavendish, "To the Reader."

69. *The Diary of Samuel Pepys,* ed. Henry B. Wheatley (New York, 1946), vol. 2, March 18, 1668.

70. Margaret Cavendish, *Grounds of Natural Philosophy divided into thirteen parts* (London, 1668), sig. A2.

71. Sarasohn, *The Natural Philosophy of Margaret Cavendish,* 27–28. See also Samuel I. Mintz, "The Duchess of Newcastle's Visit to the Royal Society," *Journal of English and German Philology* 51, no. 2 (April 1952): 168–76; and Emma Wilkins, "Margaret Cavendish and the Royal Society," *Notes and Records of the Royal Society of London* 68, no. 3 (Sept 2014): 245–60. Wilkins reconsiders Cavendish's work by placing her views within the context of the day, suggesting that Cavendish was less antagonistic toward experimentation than has been previously understood. Her treatment of Cavendish also relies less on the issue of gender than does that of Sarasohn.

72. Margaret Cavendish, epistle to the third part of the first book, in *The Worlds Olio* (London, 1655), 47–48. In the preface to the reader of Cavendish's subsequent work, *The philosophical and physical opinions* (1655), she opens with a lament about the many errors introduced into *The Worlds Olio:* "There are such gross mistakes in misplacing of chapters, and so many literal faults, as my book is much disadvantaged thereby."

73. Margaret Cavendish, *The philosophical and physical opinions written by Her Excellency the Lady Marchionesse of Newcastle* (London: Martin and Allestry, 1655).

74. Margaret Cavendish, "An Epistle to the Unbeleeving Readers in Natural Philosophy," in *The philosphical and physical opinions.*

75. Giambattista della Porta, preface to the reader, in *Natural Magik by John Baptista Porta, a Neapolitane; in twenty books* [. . .] (London: printed for Thomas Young and Samuel Speed, 1658).

76. Robert Boyle, "The Author's Preface and Declaration," in *A Defence Of the Doctrine touching the Spring and Weight Of the Air* (London, 1662).

77. Boyle, "The Author's Preface and Declaration," in *A Defence.*

78. Jean Riolan, preface to the reader, in *A sure guide, or, the best and nearest way to physick and chyrurgery,* trans. Nicholas Culpepper (London: printed by John Streater, 1671).

79. William Gilbert, "Preface to the Candid Reader," in *De Magnete,* trans. S. P. Thompson (New York: Basic Books, 1958).

80. Benedict de Spinoza, *Theological-Political Treatise,* trans. Michael Silverthorne and Jonathan Israel (Cambridge: Cambridge University Press, 2007), 258.

81. Spinoza, *Theological-Political Treatise,* 12.

82. Spinoza, preface to the reader, in *Theological-Political Treatise.*

83. John Banister, "The Epistle to the Chirurgians," in *The history of man, sucked from the sap of the most approved anatomists in this present age, compiled in most compendious*

form, and now published in English, for the utility of all godly surgeons, within this realm (London: imprinted by John Day, 1579).

84. Thomas Sherley, "To the Reader," in *A philosophical essay declaring the probable causes whence stones are produced* [. . .] (London: Printed for William Cademan, 1672).

85. Sherley, "To the Reader," in *A philosophical essay*.

86. It is also quoted by Robert Burton's in his 1621 *The Anatomy of Melancholy*, where Sherley was likely to have encountered it.

87. Anthony Ascham, *A littel treatyse of astrouomy* [*sic*] *very necessary for physike and surgerye* [. . .] (London: imprinted for Wyllyam Powell, 1550).

88. Andrew Boorde, *The Breviarie of Health* [. . .] (London, 1575).

89. Thomas Brugis, preface to the reader, in *The Marrow of Physick* (London, 1640), sig. b.

90. To get a better picture of what was printed and who it was printed for, I began with a very broad survey of records in the English Short Title Catalogue, examining works on medicine or physick published in England between 1500 and 1700. I left out items devoted expressly to anatomy, as it was clear this field remained essentially academic. Between 1500 and 1600, only nine anatomical works (in 27 editions) were published in England.

91. William Janszoon Blaeu, *A Tutor to Astronomy and Geography, or, An easy and speedy way to understand the use of both the globes, celestial and terrestrial*, trans. Joseph Moxon (London, 1654), sig. A2.

92. Thomas Willis, *The London Practice of Physick*, trans. Eugenius Philatros (London, 1692), sig. A2.

93. Nathaniel Lomax, *Launaeus redivivus: or, A true narrative of the admirable effects of Delaun's pill* (London, 1657), 4.

94. Hannah Woolley, "To All Ladies and Gentlewomen in General, Who Love the Art of Preserving and Candying," in *The ladies delight* (London, 1672).

95. Thomas Law, preface to the reader, in *Naturall experiments, or, Physicke for the poor* (London, 1657).

96. The chief source for biographical information on Salmon comes from the entry by Philip K. Wilson in *Oxford Dictionary of National Biography*. See also Craig Ashley Hanson, *The English Virtuoso: Art, Medicine and Antiquarianism in the Age of Empiricism* (Chicago: University of Chicago Press, 2009), 108–11.

97. William Salmon, *The London Almanack For the Year of Our Lord 1692* (London, 1692).

98. Average rental of shop space at the time was £17 7s 0d. The quote comes from Roger North in a family history quoted in John Timbs, *Curiosities of London: Exhibiting the Most Rare and Remarkable Objects of Interest in the Metropolis* (London: David Bogue, 1855), 469.

99. David McKitterick, " 'Ovid with a Littleton': The Cost of English Books in the Early Seventeenth Century," *Transactions of the Cambridge Bibliographical Society* 11, no. 2 (1997): 184–234.

100. Salmon, preface to the reader, in *Synopsis Medicinae* (London, 1695).

101. Elizabeth Furdell, *Publishing and Medicine in Early Modern England* (Woodbridge:

University and Suffolk Press, 2002), 43. See also Benjamin Woolley, *Heal Thyself: Nicholas Culpeper and the Seventeenth-Century Struggle to Bring Medicine to the People* (New York: Harper Collins, 2004).

102. William Salmon, *Pharmacopoeia Londinensis, or, The new London dispensatory* (London, 1678), sig. A3r.

103. *A catalogue of books printed for Thomas Basset* (London, 1672).

104. Takano Michiyo, "Richard Chiswell and his Publications in the Late 17th Century," *Yamanashi Global Studies; Bulletin of Faculty of Global Policy Management and Communications* 6 (March 2011): 74–84.

105. See *The Term Catalogues, 1668–1709*, ed. Edward Arber (London, 1903), vol. 1, xii.

106. William Salmon, preface to *Iatrica, seu, Praxis medendi, The practice of curing* (London, 1681).

107. It was printed for Henry Rhodes in Fleet Street, who had a very eclectic catalog containing not a trivial amount of erotica (including *Venus in the Cloister, or, the Nun in Her Smock*).

108. William Salmon, *The family dictionary, or, Houshold [sic] companion* (London, 1695), sig. A2r.

109. The debate between the apothecaries and the college was famously satirized by physician and poet Samuel Garth in *The Dispensary: A Poem in Six Cantos*, published in 1699.

110. *The State of Physick in London: With an Account of the Charitable Regulation made lately at the College of Physicians* [. . .] (London, 1698).

111. *The State of Physick in London*, 21.

112. William Salmon, *A Rebuke to the Authors of A Blew-Book Call'd, The State of Physick in London* (London, 1698), A2.

113. Salmon, *A Rebuke*, 22.

114. Salmon, *A Rebuke*, 24.

115. Nicholas Culpeper, *The English Physician*, ed. Michael A. Flannery (Tuscaloosa: University of Alabama Press, 2007), 102–3.

116. John Colbatch, *Four Treatises of Physick and Surgery* (London, 1698).

117. Deborah Harkness, *The Jewel House: Elizabethan London and the Scientific Revolution* (New Haven, CT: Yale University Press, 2007), chapter 6, *passim*. See also Malcolm Thick, *Sir Hugh Plat: The Search for Useful Knowledge in Early Modern London* (London: Prospect Books, 2010).

118. Hugh Plat, preface to the reader, in *The Jewell House of Art and Nature* (London: printed by Elizabeth Alsop).

119. Hugh Plat, preface to *Floraes Paradise* (London: H. Lownes, 1608).

120. Hugh Plat, preface to *Floraes Paradise*, A4.

121. Circulation of Knowledge and Learned Practices in the 17th-century Dutch Republic: A Web-based Humanities' Collaboratory on Correspondences. Stable URL: http://ckcc.huygens.knaw.nl/ (accessed September 14, 2016). See also Iordan Avramov, "Letter Writing and Management of Scientific Controversy: The Correspondence of Henry Oldenburg (1661–1677)," in *Self-Presentation and Social Identification: The Rhetoric and Pragmatics of Letter Writing in Early Modern Times*, ed. Toon van Houdt et al. (Leuven: Leuven University Press, 2002), 338.

122. Randall Ingram, "First Words and Second Thoughts: Margaret Cavendish, Humphrey Moseley, and 'the Book,'" *Journal of Medieval and Early Modern Studies* 30, no.1 (2000): 102.

CHAPTER THREE: **The Controlled Distribution of Scientific Works**

1. On books as gifts and the complex signals incumbent in the act of presentation, see Sharon Kettering, *Patrons, Brokers, and Clients in Seventeenth Century France* (Oxford: Oxford University Press, 1986); Natalie Zemon Davis, *The Gift in Sixteenth Century France* (Madison: University of Wisconsin Press, 2000); Ilana Krausman Ben-Amos, *The Culture of Giving: Informal Support and Gift-Exchange in Early Modern England* (Cambridge: Cambridge University Press, 2008); and Felicity Heal, *The Power of Gifts: Gift-Exchange in Early Modern England* (Oxford: Oxford University Press, 2014).

2. Max Caspar, *Kepler* (New York: Dover, 1993), 64.

3. Mario Biagioli, "From Ciphers to Confidentiality: Secrecy, Openness and Priority in Science," *British Journal for the History of Science* 45, no. 2 (June 2012): 12.

4. Lucien Febvre and Henri-Jean Martin, *The Coming of the Book* (London: Verso Classics, 1997), 160–61.

5. Folger Shakespeare Library, *John Ward Diaries*, vol. 16, V.a.299, f4v, digital Image #45327, accessed August 3, 2020, luna.folger.edu.

6. Douglas D. C. Chambers and David Galbraith, eds., *The Letterbooks of John Evelyn* (Toronto: University of Toronto Press, 2014), vol. 1, 949–50.

7. Stephen Gaukroger, *Descartes: An Intellectual Biography* (Oxford: Clarendon Press, 1995), 322.

8. For a general discussion of the practice see Theodorus Bögels, "Govert Basson: Printer, Bookseller, Publisher: Leiden 1612–1630" (PhD diss., Leiden University, 1992), especially chapter 3. On Descartes see Gustav Cohen, *Écrivains français en Holland* (The Hague: Martinus Nijhoff, 1921), 506.

9. Mosley, "Astronomical Books and Courtly Communication," in *Books and the Sciences in History*, ed. Marina Frasca-Spada and Nick Jardine (Cambridge: Cambridge University Press, 2000), 124. As Febvre and Martin show, however, it was not uncommon for printers to require that authors purchase a certain number of copies from a print run, thereby ensuring a certain return on the investment. See Febvre and Martin, *The Coming of the Book*, 161.

10. Publication dates are either 1598 or 1602, depending on the print run. The 1598 copies were published in Nuremberg, and the 1602 in Wendesburg. See also Mosley, "Astronomical Books and Courtly Communication," 116.

11. Mosley, "Astronomical Books and Courtly Communication," 124.

12. Domenico Bertoloni Meli, "Shadows and Deception: From Borelli's *Theoricae* to the *Saggi* of the Cimento," *British Journal for the History of Science* 31 (1998): 383; W. E. Knowles Middleton, *The Experimenters: A Study of the Accademia del Cimento* (Baltimore: Johns Hopkins Press, 1971), 78.

13. Mosley, "Astronomical Books and Courtly Communication," 124.

14. Natalie Zemon Davis, *The Gift in Sixteenth Century France*, 37.

15. Tycho Brahe, *Astronomiae Instauratae Mechanicae* (Nuremberg, 1598).

16. Huygens to Marin Mersenne, October 28, 1646, in *Oeuvres Complètes de Christiaan Huygens* (La Haye: M. Nijhoff, 1888–1950), vol. 1, 24. Hereafter cited as *OCCH*.

17. The list can be found in the *Codices Hugeniorum* 5, 73.

18. Huygens to D. Seghers, January 23, 1652, *OCCH* vol. 1, 168.

19. Huygens to Lodewijck Huygens, January 19, 1652, *OCCH* vol. 1, 167.

20. *The Philosophical Writings of Descartes*, trans. and ed. John Cottingham et al. (Cambridge: Cambridge University Press, 1991), vol. 3, 92.

21. Huygens to Fr. van Schooten, July 1, 1654, *OCCH*, vol. 1, 287.

22. *Codices Hugeniorum*, Hug 45, 2 folio, 31–63, July 1, 1654, Leiden University Archives.

23. Constantijn Huygens to J. J. Stöckar, October 13, 1654, *OCCH*, vol. 1, 298–99.

24. Constantijn Huygens to Princess Palatine Elizabeth, December 25, 1654, *OCCH*, vol. 1, 313.

25. Richard Waller, ed., preface to the reader, *The Posthumous Works of Robert Hooke* [. . .] *Containing his Cutlerian Lectures* [. . .] (printed by Sam Smith and Benjamin Walford, 1705).

26. Nicole Howard, "Rings and Anagrams: Huygens' System of Saturn," in *The Papers of the Bibliographical Society of America* 98, no. 4 (December 2004): 477–510.

27. Huygens to Boulliau, December 26, 1657, *OCCH*, vol. 3.

28. Harold Love, *Scribal Publication in Seventeenth-Century England* (New York: Oxford University Press, 1993), 44. Love's discussion of the myriad conditions which allow one to call something a "publication" (chap. 2) elaborates on these issues in important ways.

29. Huygens to J. Hevelius, March 8, 1656, *OCCH*, vol. 1, 387.

30. J. Hevelius to Huygens, June 22, 1656, *OCCH*, vol. 1, 434–35.

31. Huygens to J. Chapelain, March 1656, *OCCH*, vol. 1, 390.

32. J. Chapelain to Huygens, April 8, 1656, *OCCH*, vol. 1, 397–98. Gilles Ménage was a reputable man of letters in the period. *OCCH*, vol. 1, no. 278n4.

33. See Luciano Boschiero, *Experiment and Natural Philosophy in Seventeenth Century Tuscany* (Dordrecht: Springer, 2007), chapter 8; and W. E. Knowles Middleton, *The Experimenters: A Study of the Academia del Cimento* (Baltimore: Johns Hopkins University Press, 1971).

34. J. Chapelain to Huygens, 13 May 1660, *OCCH*, vol. 3, 80.

35. Leopold de' Medici to Huygens, 5 November 1660, *OCCH*, vol. 3, 171.

36. Descartes, *Philosophical Writings*, vol. 3, 51.

37. Descartes, *Philosophical Writings*, vol. 3, 61.

38. Descartes, *Philosophical Writings*, vol. 3, 52–53.

39. Gaukroger, *Descartes*, 323.

40. Descartes, *Philosophical Writings*, vol. 3, 150.

41. René Descartes, *Meditations on First Philosophy with Selections from the Objections and Replies*, ed. John Cottingham (Cambridge: Cambridge University Press, 1996), xlv.

42. Descartes to Gibieuf, November 11, 1640, Descartes, *Philosophical Writings*, vol. 3, 158 (italics are my own).

43. Descartes to Gibieuf, November 11, 1640, 168.

44. Christiaan Huygens, *Cosmotheros* (The Hague, 1698), 4.

45. Huygens, *Cosmotheros*, 4–5.

46. J. T. Desaguliers, preface to *A Course of Experimental Philosophy* (John Senex, 1734), vol. 1.

47. This actually varies by discipline, with contributions from the lay public more welcomed in botany and natural history, and less so in mathematics and natural philosophy. See Alice Stroup, *A Company of Scientists: Botany, Patronage and Community at the Seventeenth-Century Parisian Royal Academy of Sciences* (Berkeley: University of California Press, 1990), 182–85.

48. *OCCH*, vol. 17, 45.

49. States General to S. Coster, June 16, 1657, *OCCH*, vol. 2, 237–38.

50. Herbert H. Rowen, *John de Witt, Grand Pensionary of Holland, 1625–1672* (Princeton, NJ: Princeton University Press, 1978), especially 416–17.

51. Huygens to H. Oldenburg, May 31, 1673, in *The Correspondence of Henry Oldenburg*, ed. A. Rupert Hall and Marie Boas Hall (Madison: University of Wisconsin Press, 1969), vol. 9, 674–75. Hereafter cited as *CHO*.

52. H. Oldenburg to Huygens, June 12, 1673, *OCCH*, vol. 7, 304.

53. This letter accompanied a letter Christiaan had sent to his father, wherein he defended himself against Hooke's claims to priority of the circular pendulum. Constantijn translated it and attached these remarks before sending it to England. See Christiaan Huygens to Constantijn Huygens, father, August 17, 1674, *OCCH*, vol. 7, 390–91.

54. *Diary of Robert Hooke* (London: Taylor and Francis, 1935), 45. May 30, 1673.

55. *The Cambridge Illustrated History of Astronomy*, ed. Michael Hoskin (Cambridge: Cambridge University Press, 1996), 143.

56. *OCCH*, vol. 21, 201.

57. *OCCH*, vol. 21, 203.

58. Extant copies of the work are frequently described as having one full-page, folded engraving, as well as one *inserted*, folding, engraved plate.

59. CH "Portefeuille Musica" (Hug. 27 ff. 01v). See also Joella Yoder, *A Catalogue of the Manuscripts of Christiaan Huygens Including a Concordance with His "Oeuvres Complètes"* (Leiden: Brill, 2013).

60. See C. D. Andriesse, *Huygens: The Man behind the Principle*, trans. Sally Miedema (Cambridge: Cambridge University Press, 2005), 334.

61. Stroup, *A Company of Scientists*, 51–52. Stroup has shown that where Colbert spent an average of 87,700 livres per year on the Academy, Louvois spent just 26,484. Louvois also canceled research on the determination of longitude at sea, as well as plans to publish astronomical treatises, both of which were primary areas of study for Huygens.

62. Huygens to F.M. Le Tellier, Marquis de Louvois, *OCCH*, vol. 8, 489.

63. "L'imprimeur au lecteur," *Journal des Sçavans*, January 1665, n.p. For background, see Harcourt Brown, "The Learned Journal," *Journal of the History of Ideas* 33, no. 3 (July–Sept 1972): 365–78.

64. On the *Philosophical Transactions*, see David A. Kronick, *"Devant le Deluge" and Other Essays on Early Modern Scientific Communication* (Oxford: Scarecrow Press, 2004), chapter 10; and N. Moxham, "Fit for Print: Developing an Institutional Model of Scien-

tific Periodical Publishing in England, 1665–ca 1714," in "350 Years of Scientific Periodicals," special issue of *Notes and Records of the Royal Society of London* 69, no. 3 (September 20, 2015): 241–60.

65. On "information overload," see the introductory essay by Daniel Rosenberg, "Early Modern Information Overload," *Journal of the History of Ideas* 64, no. 1 (2003): 4. Rosenberg's is the introduction to an entire issue dedicated to this theme.

66. Thomas Broman, "Criticism and the Circulation of News: The Scholarly Press in the Late Seventeenth Century," *History of Science* 51, no. 2 (June 2013): 135.

67. "Extrait du Journal d'Angleterre contenant la manière passer le sang d'un animal dans un autre," *Journal des Sçavans* (January 31, 1667): 69–72. Discussed in Holly Tucker, *Blood Work: A Tale of Medicine and Murder in the Scientific Revolution* (New York: W. W. Norton, 2011), 68–70.

68. Tucker, *Blood Work*, 129.

69. Huygens to J.P. de la Roque, June 8, 1684, *OCCH*, vol. 8, 496–97. De la Roque had proven an important ally of Huygens in 1682 dispute with certain Amsterdam booksellers who published the Low Countries' edition of the *Journal de Sçavans*. Apparently they had inserted their own commentary in a debate between Huygens and the Abbé Catelan concerning centers of balance. De la Roque printed a harsh criticism of these booksellers in the French journal entitled, "Friponnerie de certain libraire."

70. Bayle was appointed professor of philosophy at Rotterdam in 1681, putting him in close proximity to Huygens, who was residing in The Hague. When the first issue of Bayle's journal came out, he sent a copy to Huygens, starting their correspondence.

71. See Hubert Bost, *Un "intellectuel" avant la lettre: Le journaliste Pierre Bayle; L'actualité religieuse dans les "Nouvelles de la République des Lettres" (1684–1687)* (Amsterdam: APA Holland University Press, 1994); and the review of the same by L. W. Brocolis in *Library* 18, no. 3 (1996): 266–67.

72. *Nouvelles de la République des Lettres*, 1st ed. (Amsterdam: chez Henry Desbordes, dans le Kalver-Straat, prés le Dam. 1684. Avec Privilege des Etats de Holland and Westfalia. in 12°, 1684). Bayle had informed Hugyens of his publication earlier in the year, urging him to share whatever new discoveries he had through the journal. P. Bayle to Huygens, May 29, 1684, *OCCH*, vol. 8, 490–91.

73. A. H. Laeven, *The "Acta Eruditorum" under the Editorship of Otto Mencke (1644–1707): The History of a Learned Journal between 1683 and 1707*, trans. Lynn Richards (Amsterdam: APA Holland University Press, 1990), appendix, chart of reviews and reviewers. Authors of reviews could be ascertained only from an examination of the original proof, where the writers' names were found in the margins. Pfautz wrote eighty-eight reviews between 1682 and 1707, primarily on works under the heading of "Mathematica" (under which the review of the *Astroscopia* also appeared).

74. *Philosophical Transactions* 14, no. 161 (1684): 668–70. An account of the *Astroscopia* was read at the Royal Society meeting of July 23, 1684.

75. See in J. E. Hofmann, *Leibniz in Paris* (Cambridge: Cambridge University Press, 1974), 354.

76. Huygens to P. Bayle, February 17, 1690, *OCCH*, vol. 9, 369–70.

77. Huygens to N. Fatio de Duillier, February 7, 1690, *OCCH*, vol. 9, 357–58.

78. Huygens to N. Fatio de Duillier, February 7, 1690, 357–8.

79. Huygens to N. Fatio de Duillier, February 7, 1690, 357–8. Huygens to Constantijn Huygens, brother, February 7, 1690, *OCCH*, vol. 9, 361.

80. William Albury, "Halley and the *Traité de la lumière* of Huygens: New Light on Halley's Relationship with Newton," *Isis* 62, no. 4 (Winter 1971): 450.

81. N. Fatio de Duillier to Huygens, March 6, 1690, *OCCH*, vol. 9, 384.

82. Huygens to G.W. Leibniz, February 8, 1690, *OCCH*, vol. 9, 366–68.

83. Ph. de la Hire to Huygens, 1 March 1690, *OCCH*, vol. 9, 377. This delay was likely due to the fact two editions of the work were printed, one for Huygens' personal distribution and a second—of slightly lesser quality—for sale.

84. The copies had been mistakenly seized by an official in Peronne, though neither Huygens nor de la Hire knew it at the time.

85. Ph. de la Hire to Huygens, 11 May 1690, *OCCH*, vol. 9, 419–20.

86. Andriesse, *Huygens*, 334.

87. Huygens to Ph. de la Hire, August 24, 1690, *OCCH*, vol. 9, 469.

88. R. W. Searjeantson, "Proof and Persuasion," in *The Cambridge History of Science*, vol. 3 (Cambridge: Cambridge University Press, 2006): 164–65.

89. Adrian Johns, "Miscellaneous Methods: Authors, Societies and Journals in Early Modern England," *British Journal for the History of Science* 33, no. 2 (June 2000): 159–86.

CHAPTER FOUR: **"A True and Ingenious Discovery"**

1. Victor E. Thoren, *The Lord of Uraniborg. A Biography of Tycho Brahe* (Cambridge: Cambridge University Press, 1990), 186.

2. See Joanna Kostylo, "From Gunpowder to Print: The Common Origins of Copyright and Patent," in *Privilege and Property: Essays on the History of Copyright*, ed. Ronan Deazley et al. (Cambridge: Open Book, 2010), 22

3. Cited in Jens Vellev, "Tycho Brahe's Paper Mill on Hven and N. A. Møller Nicolaisen's Excavations, 1933–1934," in *Tycho Brahe and Prague: Crossroads of European Science*, John Robert Christianson et al. (Berlin: Verlag Harri Deutsch, 2002), 334.

4. John Robert Christianson, *On Tycho's Island: Tycho Brahe, Science, and Culture in the Sixteenth Century* (Cambridge: Cambridge University Press, 2000), 136–38. See also Christianson, "Tycho Brahe in Scandinavian Scholarship," *History of Science* 36 (December 1998): 467–84. There Christianson describes the archaeological findings of N. A. Møller Nicolaisen at Brahe's site in Hven where they found the paper mill, parchment mill, and grist mill, all of which were driven by a twenty-foot water wheel. Images and further discussion of the paper mill can be found in Jens Vellev, "Tycho Brahe's Paper Mill on Hven and N. A. Møller Nicolaisen's Excavations, 1933–34," in Christianson et al., *Tycho Brahe and Prague*, 333–55.

5. Hans Ræder, Elis Strömgren, and Bengt Strömgren, eds. and trans., *Tycho Brahe's Description of His Instruments and Scientific Work as Given in "Astronomiae Instauratae Mechanica"* (Copenhagen: Det Kongelige Danske Videnskabernes Selskab, 1946), 118. For discussion of Tycho's instruments, including the press, see Jole Shackleford, "Tycho Brahe, Laboratory Design, and the Aim of Science: Reading Plans in Context," *Isis* 84, no. 2 (June 1993): 211–30.

6. Adam Mosley, *Bearing the Heavens: Tycho Brahe and the Astronomical Community of the Late Sixteenth Century* (Cambridge: Cambridge University Press, 2007), 121n24.

7. Adam Mosley, *Bearing the Heavens: Tycho Brahe and the Astronomical Community of the Late Sixteenth Century* (Cambridge: Cambridge University Press, 2007), 121.

8. On the links between artisanal knowledge and science see Pamela Long, *Openness, Secrecy, Authorship: Technical Arts and the Culture of Knowledge from Antiquity to the Renaissance* (Baltimore: Johns Hopkins University Press, 2001); Jim Bennett, "Knowing and Doing in the Sixteenth Century: What Were Instruments For?," *British Journal for the History of Science* 36, no. 2 (June 2003): 129–50; Pamela Long, *Artisan/Practitioners and the Rise of the New Sciences, 1400–1600* (Corvallis: Oregon State University Press, 2011); Pamela H. Smith, *The Body of the Artisan: Art and Experience in the Scientific Revolution* (Chicago: University of Chicago Press, 2004); and Pamela H. Smith, Amy R. W. Meyers, and Harold Cook, eds., *Ways of Making and Knowing: The Material Culture of Empirical Knowledge* (Chicago: University of Chicago Press, 2017).

9. Lesley Cormack, Steven A. Walton, and John A. Schuster, eds., *Mathematical Practitioners and the Transformation of Natural Knowledge in Early Modern Europe* (Heidelberg: Springer, 2017), 81–82.

10. *The Diary of Samuel Pepys*, ed. R. Latham and W. Matthews (London: Bell & Hyman, 1972), vol. 6, 337 (December 22, 1665). See also *The Diary of John Evelyn*, ed. E. S. de Beer (Oxford: Clarendon Press, 1955), passim.

11. D. Graham Burnett, *Descartes and the Hyperbolic Quest: Lens Making Machines and Their Significance in the Seventeenth Century* (Philadelphia: American Philosophical Society, 2005), 1–3.

12. The work, *Memorien aangaande het slijpen van glasen tot verrekijckers*, was written in 1685 and published posthumously in 1703. See Fokko Jan Dijksterhuis, *Lenses and Waves: Christiaan Huygens and the Mathematical Science of Optics in the Seventeenth Century* (New York: Kluwer Academic, 2004), 62.

13. Bennett, "Knowing and Doing in the Sixteenth Century," 133.

14. Adrian Johns, "The Past, Present, and Future of the Scientific Book," in *Books and the Sciences in History*, ed. Marina Frasca-Speda and Nick Jardine (Cambridge: Cambridge University Press, 2000), 416.

15. See "Forms and Functions of Early Modern Celestial Imagery," a special issue of the *Journal for the History of Astronomy* 41, no. 3 (August 2010): 283–424; Dániel Margócsy, *Commercial Visions: Science, Trade and Visual Culture in the Dutch Golden Age* (Chicago: University of Chicago Press, 2014), chapter 6, "Knowledge as Commodity: The Invention of Color Printing"; Sachiko Kusukawa, *Picturing the Book of Nature: Image, Text, and Argument in Sixteenth-Century Human Anatomy and Medical Botany* (Chicago: University of Chicago Press, 2012).

16. Matthijs van Otegem, "The Relationship between Word and Image in Books on Medicine in the Early Modern Period," in *Cognition and the Book Typologies of Formal Organisation of Knowledge in the Printed Book of the Early Modern Period*, ed. K. A. E. Enenkel and Wolfgang Neuber (Leiden: Brill, 2005): 603–20. Van Otegem's study of Descartes' *Treatise on Man* is especially illustrative here.

17. Robert W. Unwin, "A Provincial Man of Science at Work: Martin Lister, F.R.S., and

His Illustrators 1670–1683," *Notes and Records of the Royal Society of London* 49, no. 2 (July 1995): 211.

18. This is often, though not always, the case. For example, Descartes attempted to do his own drawings for *Dioptrique* (1637) but was dissuaded after showing some of the work to Constantijn Huygens, father of Christiaan. Huygens urged Descartes to use a professional draftsman, engraver, and woodcutter for the job, which he eventually did. See Van Otegem, "The Relationship between Word and Image," 609.

19. See Brian Ogilvie, "Image and Text in Natural History, 1500–1700," in *Images in Early Modern Science* (New York: Springer Science+Business Media, 2003), 141–66.

20. Nick Kanas, *Star Maps: History, Artistry, and Cartography*, 2nd. ed (London: Springer-Praxis Books, 2009), 163.

21. Hevelius to Oldenburg, October 11, 1667, in *The Correspondence of Henry Oldenburg*, ed. A. Rupert Hall and Marie Boas Hall (Madison: University of Wisconsin Press, 1969), vol. 3, 519. Hereafter cited as *CHO*.

22. Hevelius to Oldenburg, October 11, 1667, 519.

23. Hevelius to Oldenburg, September 19, 1668, *CHO* vol. 5, 52.

24. Oldenburg to Hevelius, August 2, 1669, *CHO* vol. 6, 169. Oldenburg's letter records his sale of nineteen copies of the *Cometographia* (seventeen copies at £30, two copies at £28), eight copies of *Selenographia* (seven at £30, one at £28), fifteen copies of *Mercurius* at £5, fourteen copies of *Mantissa* at £5, fifteen copies of *Prodromus* at £2.5, twelve copies of *Epistolae* at £2.5, two copies of *Historia Coelestis* at £30, and one copy of *Saturni Facies* at £2. Oldenburg then tallies the revenue from the sales and compares that to his own expenditures (customs charges, the price of the telescope, porters fees, etc.). The letter is a good representation of the kind of distribution a scientist could obtain through the right channels.

25. Oldenburg to Hevelius, October 28, 1668. *CHO* vol. 6, 116.

26. Hevelius to Oldenburg, December 11, 1668, *CHO* vol. 6, 244–45 (emphasis added).

27. Katherin Müller, "How to Craft Telescopic Observations in a Book: Hevelius's *Selenographia* (1647) and Its Images," *Journal for the History of Astronomy* 41, no. 3 (August 2010): 370.

28. Müller, "How to Craft Telescopic Observations in a Book," 371. Müller quotes from Hevelius' *Selengraphia, sive, Lunae Descriptio* (Gdańsk: printed by Andreas Hünefeld for the author, 1647), 219–20.

29. Matthew Hunter, "The Theory of the Impression according to Robert Hooke," in *Printed Images in Early Modern Britain: Essays in Interpretation*, ed. Michael Hunter (Surrey: Ashgate, 2010), 177–78.

30. Meghan C. Doherty, "Discovering the 'True Form': Hooke's *Micrographia* and the Visual Vocabulary of Engraved Portraits," *Notes and Records of the Royal Society of London* 66, no. 3 (September 2012): 211–34.

31. Doherty, "Discovering the 'True Form,'" 218. Doherty cites Hooke's diary entries, especially from 1672–1674.

32. Historians today admire Hevelius's printed astronomical works but also share his sadness over losing those works, and the press itself, in a fire that ravaged his observatory in 1679. Of the event he wrote to Louis XIV, "Saved by God's Mercy were . . . Kepler's

immortal Works, which I purchased from his Son, my Catalogue of Stars, my New and Improved celestial Globe, and the thirteen Volumes of my Correspondence with learned Men and the Crowned Head of all Lands . . . But the cruel Flames have consumed all the Machines and Instruments conceived by long Study and constructed, alas, at such great Cost, Consumed also the Printing Press with Letters . . . consumed, finally my Fortune and the means which God's Mercy had granted me to serve the Royal Science."

33. Ted McCormick, *William Petty: And the Ambitions of Political Arithmetic* (Oxford: Oxford University Press, 2009), 155.

34. The first discussion of his invention came in William Petty, *The Advice of W.P. to Mr. Samuel Hartlib for the Advancement of Some Particular Parts of Learning* (London, 1647), sig. A2r. The subsequent pamphlet, *A declaration Concerning the Newly Invented Art of Double Writing*, appeared in 1648. A manuscript of this pamphlet is found in Hartlib's papers.

35. Petty, *The Advice of W.P. to Mr. Samuel Hartlib*, sig. A2r.

36. Petty, *A Declaration Concerning the Newly Invented Art of Double Writing* (London, 1648), 9.

37. William Petty, *Double Writing* (London: n.p., 1648).

38. Christoph Scheiner, *Pantographice seu Ars Delineandi* (Rome, 1631). On this invention, see David Freedberg, *The Eye of the Lynx: Galileo, His Friends, and the Beginnings of Modern Natural History* (Chicago: University of Chicago Press, 2002), 393–96.

39. W. Poole, "Seventeenth-Century 'Double Writing' Schemes, and a 1676 Letter in the Phonetic Script and Real Character of John Wilkins," *Notes and Records of the Royal Society* 72, no. 1 (January 2018), https://royalsocietypublishing.org/doi/10.1098/rsnr.2017.0041.

40. M. J. Braddick and Mark Greengrass, eds., "The Letters of Sir Cheney Culpepper, 1641–1657" in *Camden Miscellany*, vol. 32, *Seventeenth-Century Political and Financial Papers* (Cambridge: Cambridge University Press, 1996), 320n12. See also Donald Francis McKenzie and Maureen Bell, eds., *A Chronology and Calendar of Documents Relating to the London Book Trade 1641–1700* (Oxford: Oxford University Press, 2005), vol. 1, 228.

41. McCormick, *William Petty*, 59nn89–90.

42. William Petty to Boyle, 21 June 1648, in *Correspondence of Robert Boyle*, ed. Michael Hunter, Antonio Clericuzio, and Laurence Principe (London: Pickering & Chatto, 2001), vol. 1, 72.

43. Oldenburg to Hartlib, July 18, 1658, *CHO*, vol. 1, 171.

44. McCormick, *William Petty*, 60n93.

45. John Rushworth, *Historical Collections*, part 4 (London, 1701), vol. 2, 1112.

46. Christopher Wren Jr., *Parentalia: or, Memoirs of the Family of Wrens* (London: 1750), 214.

47. Lena Milman, *Sir Christopher Wren* (New York: Charles Scribner's Sons, 1908), 323 (appendix F, a copy of a letter from Wren to an unknown correspondent presumed to be Wilkins). See also Adrian Tinniswood, *His Invention So Fertile: A Life of Christopher Wren* (Oxford: Oxford University Press, 2001), 44–45.

48. Milman, *Sir Christopher Wren*, 324.

49. W. Poole, "Seventeenth-Century 'Double Writing' Schemes," 6.

50. Wren, *Parentalia*, 198.

51. Thomas Sprat, *The History of the Royal Society of London, for the Improving of Natural Knowledge* (London, 1667), 316; and Robert Hooke, preface to the reader, *Micrographia* (London, 1665).

52. On woodblock engraving and etching techniques, see David Landau and Peter Parshall, *The Renaissance Print 1470–1550* (New Haven, CT: Yale University Press, 1994), 21–29; and Antony Griffiths, *Prints and Printmaking: An Introduction to the History and Techniques* (Berkeley: University of California Press, 1996), passim.

53. Griffiths, *Prints and Printmaking*, 20.

54. Griffiths, *Prints and Printmaking*, 56–59.

55. John Evelyn, *Sculptura*, 1st ed. (London: Royal Society, 1662), 146.

56. John Evelyn, *Sculptura*, 2nd ed. (London: Royal Society, 1755), 79. Note that Evelyn calls copperplate engraving "Chalcography" in his work.

57. John Balgrave, *The Mathematical Jewel, Shewing the making, and most excellent use of a singular instrument so called* [. . .] (London, 1585).

58. The original can be found in the British Library: Evelyn MS.65.

59. *Diary and Correspondence of John Evelyn, F.R.S.*, ed. William Bray (London: George Bell and Sons, 1908), vol. 3, 92. In numerous publications that reference this quote by Evelyn, the citation is given to Evelyn's correspondence as edited by De Beer, but this is an error. It is only in Bray that one finds the quote.

60. Evelyn, *Diary and Correspondence*, vol. 3, 92.

61. Evelyn, *Diary and Correspondence*, vol. 3, 147–48.

62. See Dániel Margócsy, *Commercial Visions: Science, Trade, and Visual Culture in the Dutch Golden Age* (Chicago: University of Chicago Press, 2014), 174.

63. Evelyn, dedicatory preface to Boyle, *Sculptura*, 1st ed., sig A2.

64. Evelyn, *Sculptura*, 1st ed., 133.

65. Roland Freart, Sieur de Chambray, *A parallel of the Ancient architecture with the modern*, trans. John Evelyn (London, 1664), 15.

66. William Salmon, *Polygraphice, or, The art of drawing, engraving, etching, limning, painting, washing, varnishing, colouring, and dying in three books* (London, 1672).

67. Craig Ashley Hanson, *The English Virtuoso: Art, Medicine, and Antiquarianism in the Age of Empiricism* (Chicago: University of Chicago Press, 2009), 108–20.

68. Cited in Matthew Hunter, "The Theory of Impression according to Robert Hooke," in *Printed Images in Early Modern Britain: Essays in Interpretation* (Surrey: Ashgate, 2010), 174–75.

69. Nathan Flis, "Drawing, Etching, and Experiment in Christopher Wren's Figure of the Brain," *Interdisciplinary Science Reviews* 37, no. 2 (June 2012): 152.

70. Huygens to H. Oldenburg, May 19, 1669, *CHO* vol. 5, 557. On the technical details of Huygens' printing method see Eric Kindel, "Delight of Men and Gods: Christiaan Huygens's New Method of Printing," *Journal of the Printing Historical Society* 14 (2009): 5–40.

71. Huygens to Lodewijck Huygens, May 31, 1669, in *Oeuvres Complètes de Christiaan Huygens* (La Haye: M. Nijhoff, 1888–1950), vol. 6, 442. Hereafter cited as *OCCH*. The plate is reproduced in *OCCH* vol. 6, 440.

72. Oldenburg to Huygens, May 31, 1669, *CHO* vol. 5, 583.

73. Petty, *Double Writing*, 9.

74. Huygens to Oldenburg, June 16, 1669, *CHO* vol. 6, 45.

75. *Philosophical Transactions* 97 (October 1673): 6126.

76. Oldenburg to Huygens, July 5, 1669, *CHO* vol. 6, 93.

77. Originally in *OCCH* vol. 22, 233–34. I have referenced the original and use the translation provided in Kindel, "Delight of Men and Gods," 9 (emphasis added).

78. Kindel, "Delight of Men and Gods," 9.

79. Collins to Gregory, December 24, 1670, in *The Correspondence of Isaac Newton*, ed. H. W. Turnbull et al. (Cambridge: Cambridge University Press, 1959–1977), vol. 1, 55. Hereafter cited as *CIN*.

80. Collins to Gregory, December 24, 1670, 56.

81. Collins to Gregory, December 24, 1670, 57.

82. Joseph Moxon, *Mechanick Exercises* (London, 1683), 6.

83. Matthew Hunter, *Wicked Intelligence: Visual Art and the Science of Experiment in Restoration London* (Chicago: Chicago University Press, 2013), 185–86.

CHAPTER FIVE: **Silent Midwives**

Epigraph. Epistle the Third, "To my Honor'd Friend, Dr. Charleton on His Learned and Useful Works," in *The Works of John Dryden* (London, 1808), vol. 11, 15–16.

1. Robert Darnton, "What Is the History of Books?," *Daedalus* 111, no. 3 (Summer 1982): 65–83. See also Darnton, " 'What Is the History of Books?' Revisited," *Modern Intellectual History* 4, no. 3 (October 2007): 495–508.

2. Thomas R. Adams and Nicholas Barker, "A New Model for the Study of the Book," in *A Potencie of Life: Books in Society*, ed. Nicholas Barker (London: Oak Knoll Press, 2001), 5–43.

3. On the scarcity of information about editors in this period, see Ellen Valle, "Reporting the Doings of the Curious: Authors and Editors in the *Philosophical Transactions* of the Royal Society of London," in *News Discourse in Early Modern Britain: Selected Papers of CHINED 2004*, ed. Nicholas Brownlees (Bern, Switzerland: Peter Lang AG, 2006): 71–90.

4. I have consulted Hieronymus Hornschuch, *Orthotypographia*, trans. Philip Gaskell and Patricia Bradford (Cambridge: Cambridge University Library, 1972). The original was published in Leipzig by Michael Lantzenberger.

5. Hornschuch, *Orthotypographia*, 33–34.

6. Hornschuch, *Orthotypographia*, 29.

7. For more on this see Anthony Grafton, *The Culture of Correction in Renaissance Europe* (New Haven, CT: Yale University Press, 2011), esp. chapter 1.

8. John Smith, *The Printer's Grammar* (Cambridge: Cambridge University Press, 2014), 278.

9. Eileen Bloch, "Erasmus and the Froben Press: The Making of an Editor," *Library Quarterly: Information, Community, Policy* 35, no. 2 (April 1965): 110.

10. Bloch, "Erasmus and the Froben Press," 110. In his work on early modern proofreading, Percy Simpson notes that around 1634, the King's Printers in England employed

four learned correctors, each with an MA degree. Percy Simpson, *Proof-Reading in the Sixteenth, Seventeenth and Eighteenth Centuries* (Oxford: Oxford University Press, 1935).

11. Philip Gaskell, *A New Introduction to Bibliography* (New Castle, DE: Oak Knoll Press, 1995), 111n20.

12. E. J. Kenney, *The Classical Text: Aspects of Editing in the Age of the Printed Book* (Berkeley: University of California Press, 1974), 4, 25. See also Brian Richardson, *Print Culture in Renaissance Italy: The Editor and the Vernacular Text 1470–1600* (Cambridge: Cambridge University Press, 1994), 10. Along similar lines, Adrian Johns notes that a journal editor of the seventeenth century—Jean Cornand de la Crose—described himself as a "filter of credit." Adrian Johns, *The Nature of the Book* (Chicago: University of Chicago Press), 539.

13. Richardson, *Print Culture in Renaissance Italy*, 1–2.

14. John Edwin Sandys, "The Printing of the Classics in Italy," in *Reader in the History of Books and Printing*, ed. Paul A. Winkler (Englewood, CO: Indian Head, 1978), 286.

15. Richardson, *Print Culture in Renaissance Italy*, 1. Richardson points out that some editors even had their portraits included in a work. While the literature on authorial portraits is vast, the case of an editor's images as paratext is extremely rare.

16. Richardson, *Print Culture in Renaissance Italy*, 109.

17. Simpson, *Proof-Reading in the Sixteenth, Seventeenth, and Eighteenth Centuries*, 31. See also Johns, *The Nature of the Book*, 91, 103–4. Erasmus is well known for his collaboration with printers, correcting his own and others' works. This drew the comment from Julius Caesar Scaliger in his 1531 *Oratio pro M. Tullio Cicerone contra des Erasmum* that Erasmus was a "mere corrector" of texts.

18. Simpson, *Proof-Reading in the Sixteenth, Seventeenth, and Eighteenth Centuries*, chapter 1.

19. Bloch, "Erasmus and the Froben Press," 120.

20. See Rob Iliffe, "Author-Mongering: The 'Editor' between Producer and Consumer," in *The Consumption of Culture, 1600–1800: Image, Object, Text*, ed. Ann Bermingham and John Brewer (New York: Routledge, 1995); and Adrian Johns, "Miscellaneous Methods: Authors, Societies and Journals in Early Modern England," *British Journal for the History of Science* 33 (2000): 159–86.

21. William Harvey, *Exercitatio Anatomica de Motu Cordis et Sanguinis in Animalibus* (Frankfurt: William Fitzer, 1628). On the background to Harvey's discovery, see William Keynes, *The Life of William Harvey* (Oxford: Oxford University Press, 1966); J. J. Bylebyl, *William Harvey and His Age: The Professional and Social Context of the Discovery of Circulation* (Baltimore: Johns Hopkins University Press, 1979); and Thomas Wright, *William Harvey: A Life in Circulation* (Oxford: Oxford University Press, 2012).

22. John Aubrey, *Brief Lives*, ed. Andrew Clark (Oxford: Clarendon Press, 1898), vol. 1, 300.

23. Quoted in Keynes, *The Life of William Harvey*, 322.

24. Harvey overcame his disinclination to respond to critics in 1649 when he responded to the criticism of French physician Jean Riolan. Harvey claimed Riolan was "ringleader of all anatomists of this age," whose arguments against circulation could not go unanswered. Thus, in 1649 Harvey produced a short book composed of two letters to

Riolan entitled, *Exercitatio Anatomica de Circulatione Sanguinis: Ad Joannem Riolanum Filium Parisiensem.* It was published simultaneously in Rotterdam and Cambridge. Harvey was polite to his French colleague but firm in his defense of circulation, grounding his ideas in anatomical observations.

25. Ent was originally from the Low Countries, having been educated in Rotterdam. He was elected to the Royal College of Physicians in 1639 and became president of that body in 1670. In 1645, Ent was part of an important group of physicians and natural philosophers who met at Gresham College and would later form the Royal Society. See Robert G. Frank, *Harvey and the Oxford Physiologists: A Study of Scientific Ideas* (Berkeley: University of California Press, 1980), 22–24; and Christopher Hill, "William Harvey and the Idea of Monarchy," *Past and Present* 27 (April 1964): 59–61.

26. William Munk, *The Roll of the Royal College of Physicians of London*, 2nd ed. (London: Royal College of Physicians, 1878), vol. 1, 223–27.

27. Descartes to Mersenne, February 9, 1639, *The Philosophical Writings of Descartes*, trans. and ed. John Cottingham et al. (Cambridge: Cambridge University Press, 1991), vol. 3, 134.

28. George Ent, *Apologia pro Circulatione Sanguinis: Qua Respondetur Æmilio Parisano Medico Veneto* (London, 1641). Ent's treatise, at 284 pages, was the most exhaustive defense of Harvey's circulation theory put forward in print. On the nature of Ent's defense and his natural philosophy—which differed from Harvey's—see Roger French, *William Harvey's Natural Philosophy* (Cambridge: Cambridge University Press, 1994), 168–78.

29. The story of Harvey's path to publishing *De Generation Animalium* is told by William Keynes in *The Life of William Harvey*, chapter 38.

30. William Harvey, epistle dedicatory to *Anatomical Exercitations, concerning the generation of living creatures* (London: Octavian Pulleyn, 1653).

31. Harvey, epistle dedicatory to *Anatomical Exercitations*. Subsequent quotes are from the same prefatory letter.

32. Jason McElligot, *Royalism, Print and Censorship in Revolutionary England* (Woodbridge: Boydell Press, 2007), 132. See also Henry R. Plomer, *A Dictionary of the Booksellers and Printers Who Were at Work in England, Scotland and Ireland from 1641 to 1667* (London: Bibliographical Society, 1907), 67.

33. Harvey, *Anatomical Exercitations*, 418.

34. On Harvey's political positions vis-à-vis his medical work, see Alan Shepherd, "'O Seditious Citizen of the Physicall Common-Wealth!': Harvey's Royalism and His Autopsy of Old Parr," *University of Toronto Quarterly* 65, no. 3 (Summer 1996): 482–505.

35. For information on van Schooten, see J. A. van Maanen, *Facets of Seventeenth Century Mathematics in the Netherlands* (PhD diss., Rijksuniversiteit Utrecht, 1987); Joseph Hofmann, *Frans van Schooten der jungere* (Wiesbaden: Franz Steiner Verlag, 1962); and Klaas van Berkel et al., eds., *A History of Science in the Netherlands: Survey, Themes, and Reference* (Leiden: Brill, 1999).

36. J. E. Hoffmann, "Frans van Schooten," in *The Dictionary of Scientific Biography*, ed. Charles Coulston Gillispie (New York: Charles Scribner's Sons, 1970), vol. 12, 205.

37. His illustrations were in other parts of Descartes' *Discourse on Method* as well. See van Berkel et al., *A History of Science in the Netherlands*, 373. Note, too, that *The Geometry*

first appeared in 1637 as an appendix to the *Discourse*, not as a stand-alone work. On Van Schooten's geometric diagrams, see Kirsti Andersen, *The Geometry of an Art: The History of the Mathematical Theory of Perspective from Alberti to Monge* (New York: Springer, 2007), 319–21.

38. Jan van Maanen, "Precursors of Differentiation and Integration," in *A History of Analysis*, ed. Hans Niels Jahnke (London: London Mathematical Society, 2003): 48.

39. For further information on these contributors, see (respectively) Pierre Costabel, "Florimond de Beaune, érudit et savant de Blois," *Revue d'histoire des sciences* 27 (1974), 73–75; Herbert H. Rowen, *Johann de Witt, Grand Pensionary of Holland 1625–1672* (Princeton, NJ: Princeton University Press, 1978); Karlheinz Haas, "Die mathematischen Arbeiten von Johann Hudde 1628–1704) Bürgermeister von Amsterdam," *Centaurus* 4 (1956), 235–84; Jan A. van Maanen, "Hendrick van Heuraet (1634–1660): His Life and Mathematical Work," *Facets of Seventeenth Century Mathematics in the Netherlands* (Utrecht: Drukkerij Elinkwijk BV, 1987), chapter 2.

40. *Oeuvres Complètes de Christiaan Huygens* (La Haye: M. Nijhoff, 1888–1950), vol. 14, 417. Hereafter cited as *OCCH*. In the volume I consulted, the folios were out of order; thus, one will find this particular page toward the front of the work, preceding pages 9–264.

41. The second edition of van Schooten's translation included extended commentary by him, as well as *Elements of curved lines*, by Jan de Witt. Frans van Schooten, *Geometria a Renato Descartes*, 2nd ed., 2 vols. (Amsterdam, 1659–1661). See also *Renati Des Cartes Geometria, unà cum Notis Florimondi de Beaune, in Curia Blesensi Consiliarii Regii, & Commentariis Illustrata, Operâ atque Studio Francisci à Schooten* [. . .] (Frankfurt: Frederick Knoch, 1695).

42. For a few of many examples, see Fr. van Schooten to Huygens, September 20, 1651, and Huygens to Fr. van Schooten, November 11, 1651, *OCCH*, vol. 1, 145–46.

43. Van Schooten to Huygens, September 20, 1651, *OCCH*, vol. 1, 145–46 (emphasis added).

44. Huygens to Fr. van Schooten, November 11, 1651, 156.

45. Van Schooten to Huygens, November 13, 1651, *OCCH*, vol. 1, 157.

46. Huygens to Fr. van Schooten, April 157, 1654, *OCCH*, vol 1, 279.

47. Fr. van Schooten to Huygens, April 19, 1654, *OCCH*, vol 1, 286.

48. An example of the problem is found in a 1656 letter to G. P. Roberval: "When I play against another with two dice, on the condition that I will win as soon as I will make 7 points, and that he will win it as soon as he makes 6 points; and that I give to him the dice, I demand who of the two has the advantage in this, and what." Huygens to Roberval, April 18, 1656, *OCCH*, vol 1, 404.

49. Ian Hacking, *The Emergence of Probability* (Cambridge: Cambridge University Press, 1975), 93, as well as chapters 2, 7, and 11. See also Hans Freudenthal, "Huygens' Foundations of Probability," *Historia Mathematica* 7 (1980): 113–17. Huygens' complete work on probability is found in *OCCH*, vol. 15.

50. Huygens to Guy Personne de Roberval, April 18, 1656, *OCCH*, vol. 1, 404.

51. Huygens to Fr. van Schooten, April 20, 1656, *OCCH*, vol. 1, 404–5.

52. *OCCH*, vol. 14, 5. Huygens said he had composed the work in Dutch owing to the

limitations of Latin with respect to vocabulary about probability. Once composed, however, he began translating various terms into Latin, and upon handing the mss over to van Schooten, gave him the partial list of these words. See especially p. 5n13.

53. Oldenburg to Boyle, February 25, 1659/60, in *The Correspondence of Henry Oldenburg*, ed. A. Rupert Hall and Marie Boas Hall (Madison: University of Wisconsin Press, 1969), vol. 1, 358. Hereafter cited as *CHO*.

54. Anthony Wood, *Athenae Oxonienses: An Exact History of All the Writers and Bishops Who Have Had Their Education in the University of Oxford* [. . .] (London, 1820), vol. 4, 147.

55. Biographical information on Sharrock comes from both the *Dictionary of Scientific Biography*; and Leslie Stephen, ed., *Dictionary of National Biography* (New York: Macmillan, 1887).

56. Agnes Arber, "Robert Sharrock (1630–1684): A Precursor of Nehemiah Grew (1641–1712) and an Exponent of 'Natural Law' in the Plant World," *Isis* 51, no. 1 (March 1960): 3–8.

57. Sharrock to Boyle, July 13, 1664, in *Correspondence of Robert Boyle*, ed. Michael Hunter, Antonio Clericuzio, and Laurence Principe (London: Pickering & Chatto, 2001), vol. 2, 294.

58. Robert Sharrock, *The History of the Propagation and Improvement of Vegetables by the Concurrence of Art and Nature* (Oxford, 1660).

59. The third was retitled *An improvement to the art of gardening: or, An exact history of plants*.

60. Sharrock, *The History of the Propagation and Improvement of Vegetables*, sig. A2.

61. Michael Hunter, "Boyle and the Uses of Print," in *Boyle Studies: Aspects of the Life and Thought of Robert Boyle (1627–91)* (Farnam: Ashgate, 2015).

62. Robert Boyle, preface to "Some Specimens of an Attempt to make Chymical Experiments Useful to Illustrate the Notions of the Corpuscular Philosophy," in *Certain Physiological Essays* (London, 1661), following sig. P4.

63. Karen Nipps, "Cum Privilegio: Licensing of the Press Act of 1662," *Library Quarterly* 84, no. 4 (October 2014): 494–500. See also Steven Shapin and Simon Schaffer, *Leviathan and the Air Pump: Hobbes, Boyle and the Experimental Life* (Princeton, NJ: Princeton University Press, 2011), 290–91.

64. Sharrock to Boyle, January 26, 1660, *Correspondence of Robert Boyle*, vol. 1, 399.

65. Robert Boyle, preface to *New Experiments Physico-Mechanical, Touching the Spring of the Air, and Its Effects* (Oxford, 1660), sig b2.

66. Sharrock to Boyle, April 1660, in *The Works of the Honourable Robert Boyle*, ed. Thomas Birch (London, 1744), vol. 5, 419.

67. Sharrock to Boyle, November 24, 1660, *Works*, vol. 5, 419–20.

68. Sharrock to Boyle, December 16, 1660, *Correspondence of Robert Boyle*, vol. 1, InteLex Past Masters, Full Text Humanities, http://www.nlx.com/home.

69. Sharrock to Boyle, February 21, 1661, *Correspondence of Robert Boyle*, vol. 1, InteLex Past Masters, Full Text Humanities.

70. Robert Boyle, *The Sceptical Chymist, or, Chymico-physical doubts & paradoxes* (Oxford, 1661), 442.

71. Robert Boyle, "The Publisher to the Reader," *A Defence Of the Doctrine touching the Spring and Weight Of the Air* (London, 1662).

72. Iliffe, "Author-Mongering," 175.

73. Iliffe, "Author-Mongering," 175–78.

74. This story is recounted in many places but thoroughly in I. B. Cohen, introduction to *Newton's "Principia"* (Cambridge, MA: Harvard University Press, 1971), 47. See also Cohen's introduction to his translation of Newton's *Principia* (Berkeley: University of California Press, 1999), 20–22; and D. T. Whiteside, "The Prehistory of the *Principia* from 1664–1666," in *Notes and Records of the Royal Society of London* 45 (1991): 11–61.

75. I. B. Cohen, introduction to *Newton's "Principia,"* 54.

76. Thomas Birch, *The History of the Royal Society of London for Improving of Natural Knowledge* (London, 1757), vol. 4, 347.

77. I. B. Cohen describes *De Motu* as "only one small step toward Newton's *magnum opus*," and more emphatically, "*De Motu* does not yet bear the mark of genius." Cohen, introduction to *Newton's "Principia"*, 61.

78. Cohen, introduction to *Newton's "Principia"*, 61n16. Cohen cites the *Correspondence and Papers of Edmond Halley, Preceded by an Unpublished Memoir of His Life by One of His Contemporaries* [. . .], ed. Eugene Fairfield MacPike (Oxford: Clarendon Press, 1932), 74.

79. For the edited drafts, see the University of Cambridge Digital Library, Newton Papers, MS Add.3965, https://cudl.lib.cam.ac.uk/collections/newton/1.

80. Halley to Newton, May 22, 1686, in *The Correspondence of Isaac Newton*, ed. H. W. Turnbull et al. (Cambridge: Cambridge University Press, 1959–1977), vol. 2, 431. Hereafter cited as *CIN*.

81. Halley to Newton, June 29, 1686, *CIN*, vol. 2, 443.

82. Halley to Newton, June 7, 1686, *CIN*, vol. 2, 434.

83. Halley to Newton, February 24, 1687, *CIN*, vol. 2, 469.

84. Halley to Newton, June 29, 1686, *CIN*, vol. 2, 469.

85. Birch, *The History of the Royal Society*, vol. 4, 486.

86. David A. Kornick, *"Devant le Deluge" and Other Essays on Early Modern Scientific Communication* (Oxford: Scarecrow Press, 2004), 196. See also Sachiko Kusukawa, "The Historia Piscium," *Notes and Records of the Royal Society of London* 54, no. 2 (May 2000): 179–97.

87. See Alan Cook, "Edmond Halley and Newton's 'Principia,'" *Notes and Records of the Royal Society* 45, no. 2 (Jul 1999): 129–38.

88. Newton, *Principia*, in the author's preface to the reader. See Cohen, *Newton's "Principia"*, 383.

89. Johns, *The Nature of the Book*, 464.

90. Wallis to Newton, May 30, 1695, *CIN*, vol. 4, 129. See also Niccolò Guicciardini, "John Wallis as Editor of Newton's Mathematical Works," *Notes and Records of the Royal Society* 66 (March 2012): 3–17.

91. A. R. Hall, "Newton and His Editors: The Wilkins Lecture, 1973," *Notes and Records of the Royal Society of London* 29, no. 1 (October 1974): 30.

92. Isaac Newton, *The Method of Fluxions and Infinite Series*, trans. John Colson (London, 1736).

93. John Eames, F.R.S., "A brief account [. . .] of a work entitled *The Method of Fluxions and Infinite Series* [. . .]." *Philosophical Transactions* 39 (1735–1736): 320–28.

94. Eames, "A brief account," 326.

95. Paula Findlen, "Calculations of Faith: Mathematics, Philosophy, and Sanctity in 18th-Century Italy (New Work on Maria Gaetana Agnesi)," *Historia Mathematica* 38 (2011): 249. Findlen and others also detail the unfortunate mistranslation on Colson's part, taking the Italian word *la versiera* (curve) as *l'avversiera* (she-devil), an error which resulted in references to "the witch curve" and "the witch of Agnesi" among mathematicians for some time to come.

96. Donna Maria Gaetana Agnesi, *Analytical Institutions in Four Books*, trans. John Colson (London, 1801), vol. 1, vi.

97. Advertisement by the editor, in Agnesi, *Analytical Institutions*, vi.

98. On Colson's publications see Stephen, *Dictionary of National Biography*, vol. 11, 405. For a brief description of the manuscript, see *A Catalogue of the Manuscripts Preserved in the Library of the University of Cambridge* (Cambridge: Cambridge University Press, 1857), vol. 2, 74.

99. Quoted in Findlen, "Calculations of Faith," 268.

100. Maarten Ultee, "The Republic of Letters: Learned Correspondence, 1680–1720," *Seventeenth Century* 2, no. 1 (1987): 95–112. See also James Daybell, *The Material Letter in Early Modern England* (New York: Palgrave Macmillan, 2012).

101. See Ellen Valle, "Reporting the Doings of the Curious: Authors and Editors in the *Philosophical Transactions* of the Royal Society of London," in *News Discourse in Early Modern Britain: Selected Papers of CHINED 2004* (Bern: Peter Lang AG, 2006), 71–90.

102. Translation of First Charter, granted to the President, Council, and Fellows of the Royal Society of London, by King Charles the Second, A.D. 1662, accessed April 30, 2017, https://royalsociety.org.

103. Noah Moxham and Aileen Fyfee, "The Royal Society and the Prehistory of Peer Review, 1665–1965," *Historical Journal* (November 2017): 1–27, https://doi.org/10.1017/S0018246X17000334.

104. Lilo Moessner, "News Filtering Processes in the *Philosophical Transactions*," in *English Modern News Discourse: Newspapers, Pamphlets and Scientific News Discourse*, ed. Andreas H. Jucker (Amsterdam: John Benjamins, 2009), 218–19.

105. Not until the eighteenth century did journals begin to adjust to the increasing volume of articles and the demand for editorial work beyond what a single individual could provide. In 1701 the *Journal des Sçavans* created its first editorial board, and the *Journal de Trévoux*—also formed that year—did the same. See David Kronick, "Authorship and Authority in the Scientific Periodicals," in *"Devant le Deluge,"* 106. See also Moxham and Fyfe, "The Royal Society and the Prehistory of Peer Review, 1665–1965."

106. *CHO*, vol. 1, xvii–xviii.

107. "New Pneumatical Experiments about Respiration," *Philosophical Transactions* 5, no. 62 (August 1670): 3.

108. See Shapin and Shaffer, *Leviathan and the Air Pump*.

109. *Philosophical Transactions* 1, no. 12 (May 1666): 213–14.

110. Oldenburg to Boyle, December 17, 1667, *CHO*, vol. 4, 59.

111. Johns, *The Nature of the Book*, 59–60.

CONCLUSION: **Reluctance Overcome**

1. John Locke, epistle to the reader, in *An Essay concerning Human Understanding* (London, 1689).

2. Locke, epistle to the reader, in *An Essay concerning Human Understanding*.

Index

Page numbers in *italics* refer to figures.

Académie Royale des Sciences, 37, 91, 106

Accomplisht Physician, The: authorship of, 51; preface to, 51, 52

Acta Eruditorum, 110, 112, 196n73

Adams, Thomas, 140

Agnesi, Maria Gaetana, 168

anagrams, 16, 44–47, 91, 185n131, 185n133

anatomical works, 191n90

Andriesse, Cornelis, 115

Apollonius, 2, 10

Apollonius, of Perga, 10

apothecaries, 52, 79, 83, 86

Apothecaries Guild, 83

Archytas, 104

Arnauld, Antoine, 103

Ascham, Anthony, 75

Astronomiae Instauratae Mechanica (Brahe), 93, 118

astronomical ephemera, 179n11

Astroscopia Compendiaria (Huygens): address to reader, 108; distribution of copies of, 108, 109, 111–12, 114–15; illustrations, 108; promotion of, 111–12; publication of, 107–8

Aubrey, John, 132

authorial attitudes: to accuracy of print, 33; to amount and quality of books, 26–28; to censorship, 21–23; to errors in printed works, 7; to legal ramifications, 21; to pre-paredness of readers, 14, 21, 23–26, 29, 30; to print, 14, 15, 20–21, 48, 58; to priority disputes, 20, 46, 195n53; to profit, 46, 91; taxonomy of, 20–21

authors: anonymous publications of, 30–31; authority of, 88; categories of, 58–59; control over book distribution, 90, 91–92, 93, 174; as correctors, 146; deadline pressure, 31; editors and, 12–13; engagement with readers, 9, 20, 57–58, 72–73, 74; gift giving by, 90; patronage relationships, 59, 90–91, 93–94, 101; print culture and, 10, 12, 16, 19–20, 21, 25–26, 88, 90, 174, 175; privacy issue, 31, 32, 33; professional *vs.* aristocratic, 58–59; publishers and, 16, 33, 34, 69; purchase of book copies, 193n9; response to critics, 24–25, 203n24; self-publishing practices, 12, 97, 120, 121, 122

Auzout, Adrian, 39

Bacon, Francis, 58; on academic cooperation, 21; influence of, 189n57; *Miscellany*, 31; on natural knowledge, 24; note-taking method, 65–66; praise of print, 7

Baker, Thomas, 48

Baliani, Giovanni Battista, 65

Banister, John, 74

Barberini, Francesco, 42, 102

Barker, Nicholas, 140

Barrow, Isaac, 34

Bassett, Thomas, 80

Bayle, Pierre, 112, 113, 196n70, 196n72

Beale, John, 38–39

Beaune, Florimond de, 154

Bennett, Jim, 120

Bessarion, Iohannes, 2

Biagioli, Mario, 44

Blaeu, Willem Janszoon: *A Tutor to Astronomy and Geography*, 76

Blagrave, John: *The Mathematical Jewel*, 130

Blair, Ann, 26

book reviews, 109, 111–13, 167, 196n73

book sales, 29–30

Boorde, Andrew: *The Breviarie of Health*, 75

Borelli, Giovanni, 59; *Theoricae Mediceorum Planetarum*, 93